T0321143

The Advent of Relativity

The
Advent
of Relativity

Voldemar Smilga

 **World Scientific**

NEW JERSEY · LONDON · SINGAPORE · BEIJING · SHANGHAI · HONG KONG · TAIPEI · CHENNAI · TOKYO

Published by

World Scientific Publishing Co. Pte. Ltd.
5 Toh Tuck Link, Singapore 596224
USA office: 27 Warren Street, Suite 401-402, Hackensack, NJ 07601
UK office: 57 Shelton Street, Covent Garden, London WC2H 9HE

British Library Cataloguing-in-Publication Data
A catalogue record for this book is available from the British Library.

THE ADVENT OF RELATIVITY

ISBN 978-981-123-114-8 (hardcover)
ISBN 978-981-123-115-5 (ebook for institutions)
ISBN 978-981-123-116-2 (ebook for individuals)

For any available supplementary material, please visit
https://www.worldscientific.com/worldscibooks/10.1142/12128#t=suppl

Printed in Singapore

CONTENTS

FOREWORD

During his illustrious life, Voldemar Smilga (1929–2009) wrote about 200 scientific papers on condensed matter physics, two monographs and three popular science books. Two of the popular science books were translated into English and published in Moscow in the nineteen sixties.

One of them entitled *In the Search for Beauty*, republished by World Scientific in 2019, was devoted to the historical attempts to prove the fifth Euclidean postulate, eventually leading to the discovery of non-Euclidean geometry.

The current book also narrates a dramatic story of the development of scientific ideas, this time not in geometry, but in physics. The author explains how the physicists, passing through confusions and fallacies, finally understood the nature of light and the nature of spacetime we live in.

It was first published in Russian back in 1960. The English translation by Vladimir Talmy was published by Mir publishers in Moscow in 1964, under the title *Relativity and Man*. But for this edition, I have refined the translation a bit: mostly, corrected the terminological bugs.

The witty and amusing illustrations belong to Boris Zhutovsky.

Andrei Smilga
January 2021

INTRODUCTION,

in which the author confides in the gentle reader and endeavours to explain, most didactically, why and wherefore he ever sat down to write this book

I do not know what I may appear to the world, but, to myself I seem to have been only like a boy playing on the seashore, and diverting myself in now and then finding a smoother pebble or a prettier shell than ordinary, whilst the great ocean of truth lay all undiscovered before me.

NEWTON

This book is about the special theory of relativity. A secondary schooling is adequate to understand it, but reading it calls for some mental concentration as well as an ability for mental abstraction. The unsophisticated reader may, therefore, find it difficult and boring. Nevertheless, since the discourse is abundantly interspersed with general statements and sundry examples and analogies, and insofar as most statements of fact are declared but not proved, the book can probably be classified as popular science reading.

Quite a number of popular books have been written on Einstein's theory, some of them by celebrated scientists. The author regrets that he does not rank among them. Under the circumstances, can he hope that this effort of his will compare favourably with other popular expositions of relativity theory? After all, it contains hardly an example, consideration, fact or generalisation which has not been taken, in full or in part, from some other source. More, the very presentation of the subject matter, the structure and layout of the book, are also borrowed. Still, there is one consideration which might be advanced in its favour: the author has dipped into many sources in the hope of gleaning all that is good from different works. His "creative individuality" bodies forth only in his evaluation of certain published books. In short, as they say in such cases, only the author's blunders are really his own.

This method of work is not new, but the author has adhered to it with the utmost consistency and zeal. Moreover, he thinks that the task of gathering gems of thought from different sources is both honourable and rewarding, and he will be genuinely happy if the reader finds that he has been at least partially successful.

Now a few words about why all that follows was ever set down on paper.

The style and presentation of any narrative (even if you are simply retelling somebody else's ideas in your own words) depends, naturally enough, on the narrator's own attitude to the facts he is relating.

The motivation for writing this essay was one of awe and admiration. It is this feeling of reverence for real science and real scientists, this wonder at the power of the human mind, that I have sought to convey. If you muster sufficient patience to reach the end of the book, you may get to understand better the psychology of the scientist and realise what a really wonderful thing physics is.

But now you might well ask: What has all the foregoing got to do with Einstein's theory?

Well, the theory of relativity is probably the most striking of all physical investigations. Since the author is in some ways associated with physics, he naturally considers physics to be the most wonderful of all sciences. This explains why he has chosen to write about relativity theory. But there is still another reason.

Einstein's theory has been the victim of an unhappy paradox. Its revolutionary role lies not only in its revision of purely physical concepts. Equally important is that after Einstein it is no longer possible to employ many a "self-evident" notion, term or assertion, which on closer inspection often turn out to be absolutely meaningless. At the same time, though, no other physical theory has given rise to as many absurd and meaningless notions. This applies especially to the layman's understanding of it.

Our task is to comprehend the purely physical essence of the theory, without going into problems which cannot be explained here with sufficient clarity. In particular, I have reluctantly given up the idea of discussing the general theory of relativity. You should appreciate the extent of the sacrifice, for nothing is more delightful than to argue about obscure things and thereby demonstrate one's erudition.

The incomprehension of Einstein's theory displayed by many non-physicists is, in my view, due not so much to the intricacies of the theory itself as to the fact that the fundamentals of classical Newtonian mechanics (to say nothing of the classical electromagnetic theory) are as obscure to the non-specialist as the most involved and abstract postulates of modern science. What is worse, whenever Newtonian mechanics is discussed there persists the illusion of understanding bred by the fact that most people have studied its fundamentals at school. It is probably this illusion that has given rise to the popular belief that, unlike nineteenth-century physics, modern physics is inaccessible to the uninitiated. From this follows the inevitable conclusion that "science today is more difficult than it was before".

And yet the content of Newtonian physics is no simpler (perhaps even more complex) than that of Einstein's theory. That is why, before discussing Einstein, it is important to retrace the road which led to relativity theory. Besides, genuine respect for scientists can appear only when one realises, to some extent at least, how difficult their work is.

And one final remark. In physics, no declarations can be taken for granted, even if they come from the lips of Einstein or Newton. Probably the worst affront to the memory of Albert Einstein, the greatest physicist of our time, would be to declare the validity of some postulate only because it was advanced by him. Therefore any statement in the course of our conversation — and you will probably find many new and unusual ideas even in questions ordinarily considered to be quite clear — should be accepted with caution.

It may well happen that you will feel a bit

dismayed after a closer acquaintance with physics. Physical theories seem so graceful and consummate, you are carried away by the sound, compelling logic of their authors and unwittingly fall under the spell of someone's ideas or, even worse, authority; you stop thinking and start quoting. This process is often facilitated by the subconscious conviction that "everything of consequence in science has already been done", even if you verbally protest to the contrary. The mind ceases to work and in time the habitual is more and more readily accepted as the veritable. This attitude can hardly be changed by exhortations alone.

The subject of our discourse — relativity theory — is the best example of the eternal incompleteness of science. This should be apparent when you finish the book and will be in a better position to appreciate the truth of Newton's modest appraisal of his work quoted in the epigraph.

And finally, in writing this book the author has taken freely from the works and ideas of L. I. Mandelshtam and S. I. Vavilov. It is to their memory that he humbly dedicates this effort.

CHAPTER I,

which is devoted entirely to the
one who laid the foundations

GALILEO.
THE PRINCIPLE OF RELATIVITY

*My purpose is to set forth a very new
science dealing with a very ancient subject.
There is, in nature, perhaps nothing older than
motion, concerning which the books written by
philosophers are neither few nor small; nevertheless I have discovered by experiment some
properties of it which are worth knowing and
which have not hitherto been either observed or
demonstrated.... This and other facts, not few in
number or less worth knowing, I have succeeded
in proving; and what I consider more important,
there have been opened up to this vast and most
excellent science, of which my work is merely
the beginning, ways and means by which other
minds more acute than mine will explore its
remote corners.*

GALILEO

"I, Galileo Galilei, son of the late Vincenzio
Galilei of Florence, aged seventy years, being
brought personally to judgement, and kneeling
before you, Most Eminent and Most Reverend

A rather lengthy de-
scription of Galileo's
time: the age of the
Renaissance.

7

Lord Cardinals, General Inquisitors of the Universal Christian Commonwealth against heretical depravity, having before my eyes the Holy Gospels which I touch with my own hands, swear that I have always believed, and will in future believe every article which the Holy Catholic and Apostolic Church of Rome holds, teaches, and preaches. But because I have been enjoined, by this Holy Office, altogether to abandon the false opinion which maintains that the Sun is the centre and immovable, and forbidden to hold, defend or teach, the said false doctrine in any manner, and because, after it had been signified to me that the said doctrine is repugnant to the Holy Scripture, I have written and printed a book, in which I treat of the same condemned doctrine, and adduce reasons with great force in support of the same, without giving any solution, and therefore have been judged grievously suspected of heresy; that is to say that I held and believed that the Sun is the centre of the world and immovable and that the Earth is not the centre and movable.

"I am willing to remove from the minds of Your Eminences, and of every Catholic Christian, this vehement suspicion rightly entertained towards me; therefore, with a sincere heart and unfeigned faith I abjure, curse, and detest the said errors and heresies, and generally every other error and sect contrary to the said Holy Church; and I swear that I will never more in future say, or assert anything, verbally or in writing, which may give rise to a similar suspicion of me; but that if I shall know any heretic, or any one suspected of heresy, I will denounce him to this Holy Office...."

To all appearances this formula of abjuration pronounced on June 22, 1633, by Galileo Galilei, kneeling before his judges and suitably clothed in the sackcloth of a repentant criminal, represented the sequel of his whole life. The remaining nine years he spent under what we would now call "house arrest" and in almost utter seclusion. A special interdict issued by "His Holy Fatherhood" Pope Urban VIII, "humblest of God's servants", forbade him to publish his works.

Unfortunately for "the powers that be", the interdict was violated and Galileo's principal work, the true outcome of his life, his *Dialogues on the Two New Sciences*, was published in 1638 in Holland, that "den of Protestant heresy condemned equally by God and man".

Galileo, the son of an impoverished Florentine nobleman, was born in Pisa in 1564.

In the course of the preceding century or so the cramped world of the Middle Ages had expanded tremendously. Portuguese, Spanish, and later English and Dutch ships plied the Mare Incognitum (the Unknown Sea) in quest of gold, spices, ivory, slaves and the legendary El Dorado, the country of fabulous riches. Each new voyage was a thrust into the unknown, and the navigator or pilot was at least the second, if not the first, man on board. It was his duty to steer the vessel according to the naval charts and sailing directions, calculate its course and determine its position by the stars. Sailors and, more important, merchant ship-owners (whose number grew rapidly) clamoured for better charts and instruments: people had come to realise that the success of a voyage depended on them no less than on a ship's sea-worthiness. That was one of the reasons why mathematicians and astrologers (astronomers) enjoyed the utmost respect and esteem. The more fortunate sailors sometimes acquired money, fame and titles. More often, though, they got nothing. The trading companies, however, never failed to get their profits. Trade capital, and then gradually industrial capital, began to accumulate, the bourgeoisie appeared and feudal Europe began to waken from its medieval slumber.

Neither was the Church immune from the wind of change. Catholic rites and services were much too elaborate and costly. Obviously, a "cheaper church" was required. Soon Lutheranism triumphed in Germany and Scandinavia, and Calvinism in Switzerland, Holland and England. Religious wars rocked France.

The new class needed new ideas and a new, pragmatic, science. Belief in authorities was shaken for, when put to the test, the ideas of some of the most venerable sages sanctified by the authority of Rome proved to be nonsense pure and simple. In truth, it was a "dark and evil time". Some (though very few) even began to question the existence of the Almighty himself.

As in the old tale of the magician's disciple who died at the hands of the genies he had conjured forth, the new age sparked ideas which were much more revolutionary than it could possibly digest. Besides, the Catholic Church, whose power and authority were still immense, reacted quickly and drastically to the challenge.

True, as far as scientific heresy was concerned, the reformers saw eye to eye with the Catholics, and the stake in Geneva was as efficient as the stake in Rome. Especially frequently the flames leaped up in Spain, Portugal and Italy, the most powerful strongholds of the Catholic Church.

A few words about the Church and its "scientific" methods.

The system of education, which was evolved and polished over the centuries and completely subordinated to Rome, was aimed at imbuing blind faith in authority and dogma from childhood on. The method to be followed in investigating a new phenomenon was simple enough. All one had to do was find an appropriate passage in the writings of the "Fathers of the Church". If the phenomenon was found to be in contradiction with the texts — why then, all the worse for the phenomenon!

"To such people," Galileo wrote in a letter to Kepler, "philosophy is a kind of book, like the Aeneid or the Odyssey, where the truth is to be sought, not in the universe or in nature but (I use their own words) by comparing texts!... The first philosopher of the faculty at Pisa... tried hard with logical arguments, as if they were magical incantations, to tear down and argue the new planets out of heaven!"

The method yielded its results. Think of the number of talented people who wasted their lives in interpreting some obscure passage from Thomas of Aquinas (such passages abound in the writings of the reverend fathers) or in studying some such "topical" problem as how the immaculate conception was achieved! It took an outstanding mind to escape from the dogmatic prison of scholastics. But in that case....

Tommaso Campanella (1568–1639), sociologist, philosopher and astrologer, according to the categorical conclusion of the Jesuit fathers "more venomous a serpent than either Luther or Calvin": spent 27 years in 50 prisons, seven times subjected to cruel torture.

Giordano Bruno (1548–1600), philosopher: burned at the stake in Rome.

Lucilio Vanini (1584–1619), philosopher, and diehard heretic who denied, among other things, the divinity of Christ: hanged in Toulouse. Before the execution his tongue was torn out; his body was burned and the ashes scattered to the wind.

Andreas Vesalius (1515–1564), founder of scientific anatomy: sentenced to death by the Spanish Inquisition.

Michael Servetus (1509–1553), Spanish physician: burned at the stake by Calvin in Geneva.

This terrible roll could be continued on and on. Not many of the apostates lived to die a natural death. The Jesuit fathers saw to that. The new pathways in science were probably more dangerous than the roads of the conquistadors, for they held no hope of a happy end.

Executions of heretics were so common a sight that the death of such an outstanding philosopher as Giordano Bruno, burned at the stake in the presence of a huge crowd, passed almost unnoticed by his contemporaries. There are very few records concerning his death, and at one time it was even rumoured that he had been burned in effigy and his life spared.

Most people, including even the "enlightened" classes, continued to trust implicitly in

the church and harboured the wildest super-
stitions. Scientists were no exception. The
manner in which great ideas existed side by side
with fantastic absurdities is really amazing.
People embarked in full seriousness on voyages
to discover the country of dog-headed people or
the island of sirens, whose existence was considered to be quite plausible.

As late as the end of the seventeenth century,
Fellows of the Royal Society in London would
gather to see whether a spider could escape
from a circle made of powdered rhinoceros horn.
The spider would escape and the fact was duly
recorded. Then at a following meeting Newton
might report on his work.

This is hardly a cause
for ridicule. Know-
ing nothing, scientists
were ready to check
everything. However
naive it seems to us,
this was a real scien-
tific experiment.

No new science existed as yet and, what was
probably more important, there were no new
methods: they were still in the making.

Aristotle continued to reign supreme in
authorised science. Everything capable of
arousing independent thought had long since
been expurgated from his teachings and his
works, which had been commented and inter-
preted thousands of times. Aristotle, it was
known "for sure", had posthumously received
the supreme benefaction, for the Almighty
by special injunction had relieved him of the
torments of Hell due to him as a pagan. His
works were ranked almost as high as the
writings of the "Fathers of the Church" and to
question them (even if one professed acceptance
of the truth of the Scriptures) was tantamount
to heresy. Every Christian was fully aware of
the implications of such an accusation.

It was in this stimulating atmosphere that
Galileo embarked on his studies. Education
began, quite naturally, in the monastery, and
young Galileo was accepted as a novice to a

A brief outline of Gali-
leo's life.

brotherhood. Luckily, though, his father took him home, thus terminating his clerical career. Vincenzio Galilei was probably the boy's first tutor. An educated man, musician and mathematician, and undoubtedly a talented and interesting person, he passed on to his son his devotion to science and sceptical regard for authorities.

An interesting detail: Galilei Sr. is the author of a treatise on old and new music, written in the form of a dialogue (subsequently his son's favourite literary form), in which he makes some caustic remarks about the practice of quoting authorities as the ultimate argument in scientific debates. In those days such sentiments were seditious indeed.

At his father's insistence, young Galileo began to study medicine at the University of Pisa. He fared poorly, however; besides, the Aristotelian philosophy, which he had studied most assiduously, began to arouse serious doubts in his mind. Actually, though, Galileo had no intention of becoming either a physician or a physicist. His desire was to follow the profession of a painter. But Galilei Sr. intervened again and Galileo, acting on his advice, opened the books of Archimedes and Euclid. Very soon he forgot his momentary artistic dreams, and physics became the vocation of his life.

There is a fine portrait of Galileo in old age. Looking down from the canvas is not a scriptural sage nor a repentant martyr of the Inquisition but the wise, imperious and stern countenance of a person who has lived a hard and toilsome life devoted to one absorbing idea.

Galileo soon found out that a scientist without a patron was as helpless as a sailor without a compass. His first patron was the Marchese Guidubaldo del Monte, himself a scientist of no mean ability who sincerely admired the young man's talent. Later on patrons were to be procured by intrigue and humiliating flattery.

Early in his career Galileo received an objective lesson at Pisa which taught him to keep his thoughts to himself. First he aroused the universal indignation of "learned circles" by his criticisms of Aristotle; then he was forced to resign his professorship for a scathing report on an absurd project drawn up by one of the many bastards of his sovereign, Cosmo de Medici I.

For eighteen years after that Galileo held a chair at Padua. His views had already taken shape: he knew for sure the insolvency of Aristotelian physics and the truth of the Copernican doctrine. He had practically all the arguments at his finger-tips, but thirty years elapsed before his famous *Dialogue on the Two Great Systems of the World* appeared. Meanwhile he continued to lecture on the Ptolemaic system. Only his friends knew about his work. He wrote to Kepler (in 1597!):

"I have as yet read nothing beyond the preface of your book, from which, however, I catch a glimpse of your meaning, and feel great joy on meeting with so powerful an associate in the pursuit of the truth, and, consequently, such a friend to truth itself; for it is deplorable that there should be so few who care about truth, and who do not persist in their perverse mode

A letter worth reading

of philosophising. But as this is not the fit time for lamenting the melancholy condition of our times, but for congratulating you on your elegant discoveries in confirmation of the truth, I shall only add a promise to peruse your book dispassionately, and with the conviction that I shall find in it much to admire.

"This I shall do the more willingly because many years ago I became a convert to the opinions of Copernicus and by this theory have succeeded in explaining many phenomena which on the contrary hypothesis are altogether inexplicable. I have arranged many arguments and confutations of the opposite opinions, which, however, I have not yet dared to publish, fearing the fate of our master, Copernicus, who, although he has earned immortal fame among a few, yet by an infinite number (for so only can the number of fools be measured) is hissed and derided. If there were many such as you I would venture to publish my speculations, but since that is not so I shall take time to consider of it."

Now and then, however, Galileo would lose control of himself. In 1604 he had a fierce argument with scholastics concerning a new star, whose appearance contradicted the Aristotelian doctrine of the unchangeability of the starry sphere. Other questionable statements of his aroused suspicion, as did his lectures, delivered not in the traditional Latin but in vernacular Italian.

Galileo was quite aware that his character of a well-behaved Catholic was somewhat tainted, and he discreetly sought high-ranking patrons. Among his pupils we find the Grand Duke of Tuscany, who in turn appointed him "philosopher and mathematician extraordinary"; among his friends is Cardinal Barberini (the future Pope Urban VIII), to whom he dedicated many of his works. His life is filled with intrigues, priority squabbles and the exasperation of foolish opposition and lack of understanding.

"What do you say of the leading philosophers here," he wrote to Kepler, "to whom I have offered a thousand times of my own accord to show my studies, but who, with the lazy obstinacy of a serpent who has eaten his fill, have never consented to look at the planets, or Moon, or telescope?..."

He had to be constantly on his guard, but he worked indefatigably till the last days of his life. Seriously ill, exhausted physically and morally, forced to recant his beliefs and condemned, he completed his *Dialogue*, the work of his life and, half-blind (!), continued his astronomical observations. "Though I am silent, I am not quite idle," he wrote with characteristic reserve.

Decades of humiliating struggle against ignoramuses left their imprint on Galileo. He grew short-tempered, uncommunicative and morose; often cynical and derisive, he held few people in esteem and developed a highly uncomplimentary notion of the human race: "The number of thick skulls," he wrote, "is infinite."

In time he learned to evade the fashionable small talk about science with titled dilettantes and debates with countless "mateologists", as he called them.

But he was not completely alone. The close circle of friends and pupils who idolised their teacher constituted a promised haven in an ocean of ignorance. In their company he was neither the morose old man nor the smooth-tongued courtier. He was himself, a humanist, a daring and profound thinker and forever and always a great physicist in love with his science.

His pupils, incidentally, later did him a poor service. Their adoration was so great that they embellished his biography to such an extent that much of their information can hardly be trusted.

Galileo's interests were extensive; he was a brilliant scholar of ancient and modern art, and in his free time he composed sonnets. His books abound in references to poets and in witty examples; among other things, he can be regarded as the originator of a new genre in literature, popular science. In his work, however, there was no place for the poet: he dealt with facts and facts alone, investigating them carefully and without prejudice, always restraining the flight of fancy. Years would pass before he formulated his conclusions.

The very antithesis of Giordano Bruno with his sparkling imagination, with whom he is often compared, Galileo adhered to the same faith of Truth and, I am sure, possessed no less personal courage.

Probably the best estimate of Galileo was given by Lagrange: "The discovery of Jupiter's satellites, of the phases of Venus, of the sunspots, etc., required only a telescope and assiduity; but it required an extraordinary genius to unravel the laws of nature in phenomena which one has always under the eye, but the explanation of which, nevertheless, had always escaped the philosophers."

It is not our purpose to go into all, or even nearly all, of Galileo's work: just to list his researches and findings would occupy too much space. Our interest lies in one of his investigations, perhaps the most important of all: his study of the laws of motion, a study which to this day causes scientists to wonder at the greatness of his genius.

A sort of introduction to the second half of the chapter.

There will be only one more small digression before we get down to business.

The genius of Galileo can best be comprehended in comparison with Einstein. The fundamental concepts of Einstein's theory are no more involved or paradoxical than the notions advanced by Galileo. Yet most people find Galileo's ideas simple and understandable, while Einstein's theories require a great deal of brainwork. This is only natural. From childhood on we have been brought up in the spirit of classical physics and, as a rule, we find it difficult to change habitual notions.

Speaking of habitual notions, do you find it easy to believe (mind you, believe!) that the air presses on each square inch of your body with a force of 14.22 pounds?

It would be best, therefore, to forget all we know and simply draw conclusions from the facts of "the great book of Nature", to use Galileo's words.

And so, motion. Motion is undoubtedly caused by force. A cart won't move without a horse. Four horses draw it faster than two. Ergo, the greater the acting force the greater the speed. If the harness breaks the cart stops. We know this from daily experience. Hence, to maintain velocity, a force must continuously be applied. For a body to be in uniform motion in a straight line the force has to be uniform in magnitude and direction.

Attention! A minor mystification.

Don't be in a hurry to shrug your shoulders. You have, of course, immediately noticed the fallacy of Aristotle's theory of motion. But this is because you were taught the laws of mechanics at school: you received them cut-and-dried, so to say.

And yet I am almost sure that you have been deluded. Because, if I am right, you have not stopped to analyse the meaning of the concepts "uniform, rectilinear motion" and "force of uniform magnitude and direction". By using

these words we have either said very much or nothing at all. In fact, an analysis of these concepts should serve to shape one's views on space and time. You might object that space, time and force are self-evident categories and concepts which require no further definition. But remember that some of the greatest mistakes in science were made because things were assumed to be self-evident.

A rather important edifying remark.

Anticipating our narrative, I should like to point out that it must be put down to Einstein's credit that he demonstrated that even at the close of the nineteenth century physicists had no clear idea of such a "self-evident" concept as time. However, time is not our concern, for the time being, at least.

How to refute Aristotle? The proof of the pudding is the eating, and only experiment can disprove (or prove) a physical theory. Galileo was the first man in medieval Europe to realise this with utmost depth and clarity.

The idea, of course, had been expressed before. To the semi-legendary Paracelsus*

belongs the statement that "theory not confirmed by facts is like a saint who has performed no miracles". But Galileo was the first, in physics at least, who made the investigation and analysis of experiments and practical findings a firm principle in his work. Thanks to this he was able to pinpoint Aristotle's error. Galileo had no faith in words, his faith lay in experiments.

It did not take him long to establish that in equal times a body falls through increasing spaces and that a constant force — the body's weight — produces uniformly accelerated motion (though he had no clear notion of force).

Paracelsus, born Theophrastus Bombastus von Hohenheim, a famous physician and alchemist of the Middle Ages.

In order to measure short time intervals Galileo built an ingenious new water clock. Again and again he had to overcome purely practical difficulties of this kind. No one preceded him and he had to tread on entirely new ground. Still this part of his work required only great diligence and ingenuity. Much more difficult was to draw the unexpected and far from apparent conclusion: if not for the resistance of the air, "all bodies would fall similarly, that is, with the same speed from equal heights..., moving with uniform acceleration, such that in equal times their speed increases by the same value". In other words, Galileo noted that the speed of a falling body changes as a function of time according to the law so familiar today to any schoolboy:

$$V = gt.$$

Speaking of Galileo as a theoretical physicist, two points should be noted.

Firstly, he never cast about in endless search for the causes of a phenomenon, as was so characteristic of the Aristotelian school. (Actually, his was Newton's method of principles, which we shall discuss further on.) Galileo knew nothing of the law of gravitation, he was virtually ignorant of the meaning of "force", he did not know *why* the Earth attracts bodies and accelerates them all equally. His main interest lay in the manifestations of various phenomena. He sought the answers to these questions in experimental analysis.

Secondly, Galileo's wonderful intuition enabled him, in analysing phenomena, to assess and cast aside the secondary and non-characteristic in favour of the fundamental.

Thus, in studying falling bodies, he took into account not only the air drag but also the effect due to Archimedes' law. He pointed out that when a body falls through a medium it is not the total weight but its excess over the weight of the displaced fluid or medium that acts upon it.

The transition from experiment to theoretical generalisation is usually the most difficult part of any research. The physicist never studies a phenomenon in its "pure state". He can be likened to a photographer studying a picture on which several negatives have been superimposed. There are hundreds of examples on record of scientists overlooking discoveries only because they failed to grasp the significance of their observations. It was in the analysis of results that Galileo's talent displayed itself to the fullest.

The formidability of his task will be realised if we recall that Aristotle's theories of motion had remained unchallenged for some two thousand years.

In studying the laws of fall, Galileo proceeded naturally from vertical fall to motion on an inclined plane. He found that acceleration is constant with time and is the less the smaller the angle of inclination. In the limiting case of a horizontal plane, he claimed, a body will move without acceleration. The force that causes a body to move down an inclined plane is its weight. Furthermore, Galileo realised that motion on an inclined plane is caused not by the whole weight but by a portion of it, that portion being the smaller the less the slope.

Remember again that Galileo had no idea as to why a body falls to the Earth. Moreover, he had no clear understanding of the concept of

Some results of Galileo's investigations.

force, no formula connecting force and acceleration, and he had not read Newton's *Principia*, which would appear only 45 years after his death.

Yet Galileo's intuition led him to the conclusion: A body projected along a smooth horizontal plane would, if all resistances and external

impediments were removed, continue to move uniformly along that horizontal plane forever.

Thus, if there is no impressed force, velocity is constant. Any velocity once imparted to a moving body is maintained as long as the external causes of acceleration or retardation are removed.

Figuratively speaking, this formula sets the whole of mechanics from its head back to its feet.

The first formulation of a proposition which brings to mind the inertia law.

The respective theories of Galileo and Aristotle deserve careful comparison. Aristotle, it will be recalled, held that a constant force was necessary to maintain a constant velocity. Galileo's views were diametrically the opposite: velocity, he declared, is constant only if no forces whatsoever are made to act on a body.

This was not an elaboration of an old theory, not a question of extending or restricting its field of application. Far from it! This meant the rejection of the whole of Aristotelian mechanics.

Such situations occur very rarely in science, and they are usually found in its infancy. As a rule, a discoverer has some point of departure, some landmarks on the road he is following. Only the pioneers have nothing to take from their predecessors and therefore have to start from scratch.

Galileo is one of the founding fathers of physics. It fell to him to lay the foundations on which Newton later erected his system of mechanics. Much remained obscure to Galileo. Often he was mistaken and took the wrong path. This was hardly surprising, and Galileo himself realised better than anyone else both the significance and the failings of his work (recall his words quoted in the epigraph).

We find in his works many statements which might suggest that he was well aware of both the first and second laws of mechanics and that hence Newton was to some extent merely a populariser of his ideas. We should not, however, overestimate Galileo's contribution. In fact, neither he nor any other predecessor of Newton had a clear understanding of the first law of mechanics, which Galileo, it would seem, had formulated so clearly. Only in Newton's works do the laws of mechanics acquire that precise, polished form in which we know them (though, as we shall see later, Newton too erred). This evaluation of Galileo's work might seem a bit restrained. A detailed analysis, however, would lead us into deep waters, so let us proceed further.

If we follow up Galileo's reasoning, we observe a definite similarity between a body at rest and in uniform motion. A ship ploughing the high seas with constant speed and a ship at anchor are equally free of "external causes

of acceleration or retardation". Moreover, the Earth itself might be at rest or in uniform motion through space, and in either case there would be no "external causes".

But if that is the case, then maybe (maybe!) the physical processes taking place on a uniformly moving body, such as the Earth (if it is in uniform motion, of course), must be the same as if that body were at rest?

Galileo formulates this idea as well. Yes, he says, from the point of view of mechanics there is absolutely no difference between a body at rest and a body in uniform motion.

Here it is, Galileo's relativity principle!

If the laws of mechanics are valid in any frame, then they are valid in any other frame in uniform motion in a straight line relative to the original one.

This is Galileo's relativity principle, one of the most wonderful and remarkable laws of nature.

At this point several remarks are called for.

First, the observant reader will probably have noticed that we have not yet defined motion, therefore our reasoning about bodies in motion and at rest is, to say the least, groundless. Further on the concept of mechanical motion — which is not so simple as might be expected — will be examined in detail. At this stage we shall continue to follow in Galileo's footsteps, and Galileo, strange as it may seem, had no clear understanding of mechanical motion.

Secondly, later on, when Newton's laws will have been formulated, we shall see how the Galilean relativity principle can be inferred from them. At present we shall note that the inertia law is, by itself, insufficient to affirm the principle of relativity. True, we have just linked the inertia law and the relativity principle in Galileo's mind, as it were, but this was merely a literary device. Actually, although the inertia law and the relativity principle are closely connected, Galileo had rather guessed his principle, "espying" it from nature without associating it with the inertia law. This can be observed on reading the portion of his *Dialogue* where, in effect, he expounds the relativity principle. (It goes without saying that Galileo himself had never formulated the "Galilean relativity principle" as we know it today.)

Galileo's *Dialogue*, which dealt the Ptolemaic system its death blow, is a masterpiece both in content and in form. Though enjoined to abstain from promulgating the Copernican system, Galileo with his extensive connections was able to secure the publication of a book discussing that "heretical" system. He doesn't defend Copernicus. For all appearances he merely sets forth the facts of the case "Ptolemy versus Copernicus". Outwardly the author does not take sides and no conclusions are drawn. Two scholars, adherent of Copernicus and Ptolemy, argue between themselves. Galileo merely sets forth their reasoning. The reader is free to decide whose arguments are more convincing. Acting in the capacity of an impartial judge, Galileo analyses attempts to refute Copernicus with the laws of Aristotelian mechanics.

If any mechanical experiment conducted on Earth could disprove its motion around the Sun and prove it to be at rest, the argument would have been settled. In his *Dialogue* Galileo offers what seem to be sound objections to Copernicus.

If the Earth is moving, then a stone dropped from a tower should be deflected, as its fall is directed towards the centre of the Earth, but during the time of fall the Earth slips away from beneath it. A shell shot vertically up should, for the same reason, fall far away from the gun. A shell shot to the west should fly much farther than one shot to the east, as the Earth's motion, if it exists, must carry the cannon to the east, "away from the shell" in the first case, "in its wake" in the second. Clouds and birds should trail behind the Earth, etc. Daily experience, however, tells us that all this is not so. Hence the Earth is at rest?!

Incidentally, Galileo employs a very subtle method of reasoning. The arguments against the hypothesis of the Earth's motion are advanced by Salviatus, the Copernican adherent, whereas Simplicius, the Aristotelian, listens with delight and agrees with him. Then, after showing that he understands Aristotle better than the Aristotelians themselves and expounding seemingly irrefutable arguments in his favour, Salviatus Galileo makes a volte-face.

He draws an analogy between the Earth and a uniformly moving ship. All bodies on board a ship behave as if it were at rest: a stone dropped from the mast hits the deck at its base; a ball travels the same distance whether thrown, with the same force, towards the bow or the stern. No experiment carried out in a uniformly moving ship can establish whether it is in fact moving or not. Hence, *no experiment on Earth can tell us whether it is at rest or travelling through space with tremendous speed and revolving at the same time about its axis.*

This proposition of Galileo's is, of course, erroneous.

You may well shrug your shoulders and voice opposition: scores of experiments carried out on Earth can visually demonstrate its diurnal motion. Suffice it to recall Foucault's pendulum or that a stone dropped from a tower does, in fact, deviate to the east.† In our time such

---

†To be more precise, a falling body deviates to the southeast (in the Northern hemisphere), but the southerly deviation is very small as compared with the easterly.

The relativity principle is discussed in detail in Chapter V.

At this stage one might ask, what is meant by "a body in uniform motion" or "a rotating body"? Answers to these questions can be found in Chapters IV and V.

ignorance is impermissible, for these facts are known to any secondary school-leaver. The author, too, is well aware of them, but not so Galileo.

In formulating the principle of relativity Galileo failed to realise that it holds good only for uniform rectilinear motion — and he applied it to rotational motion. Today we know that mechanical phenomena in a rotating body take place in a different manner than in a body at rest or in uniform rectilinear motion.

Uniform motion along a circular path can be detected by the centrifugal forces that are developed, and therefore it can be readily distinguished from the state of rest or uniform rectilinear motion. But this truism was unknown to Galileo. Even in his mistakes, however, he was nearer to the truth than Aristotle.

A person with a secondary education will readily detect the erroneous and obscure propositions in Galileo's work, but it would take a man of equal stature to obtain results of the same magnitude as he did.

That is all about Galileo. Later on, armed and fortified with Newton's laws, we shall return to the problems raised in this chapter. We shall formulate the basic principles of mechanics. Then, after finally achieving complete clarity, we will dwell on the points which escaped Newton, on his errors.

We shall see further on how the study of another seemingly unrelated branch of physics, electromagnetic oscillation, led in time to the need for a complete revision of our notions of space and time. Physicists were driven to investigate the question: "What are time and space?"

Some day new revolutions will probably take place in physics, and several generations from now, our notions will seem as naive as some of Galileo's ideas seem to us. But physicists of all ages will always revere Galileo as the first man to realise that new ideas must be sought in "the great book of Nature", relying on facts alone.

In conclusion I should like to cite an example which offers a fine characteristic of the precision and honesty of Galileo's scientific reasoning. You may probably have heard that it was Giordano Bruno who first advanced the hypothesis of the infinity of the universe. This is not quite so. The question of the infinity of the universe seems to have interested Galileo, who returned to it several times and probably went further than Bruno.

And last.

In the absence of experimental data proving or disproving the finity or infinity of the universe, Galileo wrote: "I hesitate to declare which of the two propositions is correct, though my own conclusions incline me in favour of the infinity of the world."

CHAPTER II,

which touches very briefly on the life and character of Newton. Towards the end the reader finds out about the method of principles

NEWTON. MECHANICS (the method)

To derive two or three general principles of motion from phenomena, and afterwards to tell us how the properties and actions of all corporeal things follow from those manifest principles, would be a very great step in philosophy, though the causes of those principles were not yet discovered.

NEWTON

It is traditional to venerate Newton. The stock of enthusiastic epithets, comparisons and hyperbolas applied to him was all but exhausted by his contemporaries. Subsequent generations could only repeat them: which they do to this day without fear of wearying humanity.

Everyone reveres Newton: scientists who can really appreciate the import of his work, and laymen whose notions of it are rather vague but who know that to revere him is the correct thing to do.

The best way for the uninitiated to fully appreciate Newton's genius and scope would be to read Academician Vavilov's excellent book, *Newton*, probably the best work on the history of physics ever to appear in Russian. It would probably leave you for the rest of your life with a feeling of plain childish awe. I personally think that one can rever only what one understands, but Newton's creative genius defies understanding.

The life and career of Sir Isaac Newton was not in the least spectacular and, outwardly at least, followed the pattern of many a respectable English gentleman, often hard-working and gifted, who had pulled himself up by his bootstraps, devoted to his God and his King and firmly convinced that there was no land better than Merrie Olde England with her fine traditions and that it was the paramount duty of every Englishman to work for her (and incidentally his own) prosperity.

Newton was born in 1642, the year Galileo died, into a family of very ordinary means. He must have been fortunate to have a good schoolmaster who, it seems, was a cultured and educated man. In 1661, young Newton entered Cambridge University's Trinity College. Eight years later, in 1669, he became a professor of mathematics at Cambridge (no exceptional career in those days, when it took much less time than now to climb the ladder from student to professor). A serious and reserved youth, he enjoyed the respect of those about him, but even the few close friends he had were hardly aware of the fact that he had already invented his method of "fluxions", or calculus, drawn up a comprehensive programme of further

A short biography of Isaac Newton.

researches and formulated revolutionary ideas and findings in mechanics, optics and gravitational theory. By then he had already carried out his experiments on breaking down sunlight into different colours, so familiar today to any schoolboy, and formulated the law according to which gravitation reduces with distance.

All this he carried out in a matter of two years (1665–67), during which he also developed many experimental techniques, including the manufacture of telescopes, the most precise instruments of the time.

Newton was in no hurry to publish his works. Even more than Galileo or any other celebrated scientist, was he particular about his findings, and he never made them public until he was quite sure of their finality and unquestionable accuracy.

Like Galileo before him, Newton owed his career to the telescope. His invention of the reflector telescope earned him membership of the Royal Society, whose president he was later to become. He was elected a Fellow on January 11, 1672. On February 6 of the same year he already presented his famous *New Theory about Light and Colours*, which, as Academician Vavilov writes, "for the first time demonstrated to the world the possibilities of experimental physics, if properly approached". And ever since the world has never ceased to wonder at the scope and quality of his work.

His social standing mounted accordingly. In 1686–89 he sat in Parliament as Member for Cambridge. True, wicked tongues claim that he only took the floor once — to ask that the windows be shut because of "the foul odour coming from the Thames", Newton, however, was not the unworldly scientist that some of his biographers make him out to be. In any case, when he was appointed Warden (1696), then Master (1699), of the Mint, he successfully coped with his difficult job, which called for considerable administrative abilities, and superintended a complete recoinage.

He was knighted in 1705 and became *Sir* Isaac Newton, honoured officially and unofficially, by friends and foes alike, as the greatest "natural philosopher" of his time. His theological works were also highly acclaimed, even though he frequently deviated from canonical dogma. It is difficult to imagine Newton wasting his time on theological problems, but the fact remains that he was a very religious man and that he probably actually regarded his scientific work as his contribution towards the comprehension of God's Providence. True, in his time there was nothing extraordinary in such a combination of physical and theological researches, but to us his religious studies must seem incongruous.

An interesting aside: Newton's theological views also influenced subsequent generations.

In later years administrative and social duties frequently distracted Newton from his work. Moreover, age and the overstrain of the intense work of his earlier years began to tell. Still, he continued his researches and even his experiments. Most of his time, however, was devoted to polishing his earlier findings, especially the main work of his life, *Philosophiae Naturalis Principia Mathematica (Mathematical Principles of Natural Philosophy)*. Published in 1687, the *Principia*, as it is commonly known, expounds the theory of gravitation and planetary motion and formulates the fundamental laws of mechanics, which remained intact till Einstein's time.

And so, mechanics. First, the method.

"I frame no hypotheses," Newton writes. "For whatever is not deduced from the phenomena is to be called a hypothesis; and hypotheses, whether metaphysical or physical, whether of occult qualities or mechanical, have no place in experimental philosophy."

Hence, a hypothesis is not the result of experiment. A hypothesis may be formulated by intuitive reasoning on the basis of certain analogies; later it is balanced against the available facts. Thus, the atomic structure of matter remained a hypothesis until fairly recently.

Often a hypothesis may totally collapse under pressure of facts; scores of years may pass before such facts appear, as happened in the case of the Kant–Laplace hypothesis of the origin of the solar system.

Principles, on the other hand, are formulated on the basis of a thorough analysis of experimental data. Principles cannot be proved by speculative reasoning and they must necessarily rest on the firm foundation of experiment. Therefore, in one form or another, the principles remain forever, though subsequent investigation may restrict their applications or reveal that some principle is rather an approximate than an absolute statement of fact.

Examples of principles are the axioms of Euclidean geometry, the Newtonian laws of motion and universal gravitation and the laws of conservation.

Thus, when we propound a hypothesis we should be prepared for the possibility of new

The next few pages are devoted to Newton's method. They can safely be skipped. But they can equally safely be read, with some benefit to the reader, I hope.

facts refuting it completely. When we formulate a principle we can be sure that, even though it may prove to be only approximately correct and its field of application may become narrower than originally assumed, it will nevertheless remain in science in one form or another.

Still, when we come to think of it, the distinction between a principle and a hypothesis appears to be rather a notional one: after all, a hypothesis, too, must agree with experimental data and be based on experience. On the other hand, no one is safeguarded against wrong conclusions in analysing an experiment, and hence against formulating an erroneous principle which will later be refuted by facts.

Our task, however, is not to provide ideal definitions (hardly a rewarding undertaking at best), but to understand Newton's method of principles.

Let us approach the problem from another aspect. Consider what Newton has to say: "To derive two or three general principles of motion from phenomena, and afterwards to tell us how the properties and actions of all corporeal things follow from those manifest principles, would be a very great step in philosophy, though the causes of those principles were not yet discovered." The last words, I think, formulate the gist of the method of principles and its fundamental difference from the method of hypotheses. In his studies Newton deliberately refused to go into explanations concerning the causes and nature of phenomena or the properties of matter that underlie the general laws derived from observations. He is content to formulate those general laws.

An excellent illustration is the law of gravitation. What does Newton have to say about the nature of gravitation? What theoretical considerations are there to confirm that the force of interaction of two bodies is proportional to the product of their masses and inversely proportional to the square of the distance between them? None at all. Newton has no idea as to why the gravitational law is such. Moreover, he doesn't even care. He is content to have formulated the law on the basis of observations.

There is another way of scientific research. Having established, for example, the law of gravitation one might set forth propositions concerning the nature of gravity and draw up theoretical formulas to support the law. More, without even knowing the law one might propound no end of hypotheses concerning the nature of gravity.

The physics of hypotheses, the method of hypotheses, consists in that the scientist seeks to penetrate deeper into the nature of a phenomenon than the available facts allow. In this he naturally has to make bold — and often erroneous — surmises.

It might seem that the method of hypotheses is more interesting and ingenious than the method of principles and that fundamental science should prefer the former. But any contrasting of the two methods would be of little or no value. Both methods are employed in science and, we shall see later on, Newton himself often resorted to the method of hypotheses, even though he had good reason to be wary of them.

The method of principles, or the inductive method, as it is often called, was practically non-existent before Newton's time.* Hypotheses reigned supreme in the scientific world. Celebrated scientists and nonentities and semi-literate ignoramuses propounded whole systems designed neither more nor less than to explain all and every known natural phenomenon. Hypothetical physics was inherited from the Greeks, enthusiasts of speculative reasoning and conjecture. The efforts of the generation immediately preceding Newton prepared the ground for new methods of work.

It required courage and clear thinking to evade the temptations of hypothetical physics and to make the dry, sober, unfanciful method of principles the basis of work.

Academician Vavilov is probably right when he says that the secret of Newton's immortality lies in his choice of method. And now that we know the architect's style, let us examine the structure erected by him.

---

*Though we might say that in this, too, Galileo had in some respects anticipated Newton.

CHAPTER III,

the longest and probably most difficult chapter in the book. It discusses the theory of measurement in physics

NEWTON. MECHANICS
(analysis of basic concepts: length and time)

Get to the root!

KOZMA PRUTKOV

The laws of mechanics are as customary and natural to us as, say, electric light. We leave school with a firm belief in their absolute truth and irreproachable explicitness. There is hardly a schoolboy who doubts that, if anything, he knows and understands Newton's laws. But is that really so?

An entirely unpedagogical remark which young readers had best overlook.

A closer examination of the facts immediately reveals that this optimistic notion is due to sweet childish innocence. This may not be so strange since, after all, what can one expect of a schoolboy?

More strange is that up till the close of the nineteenth century learned men whose names

40

have deservedly been inscribed in gold in the annals of science failed to notice among the basic propositions of Newtonian mechanics a number of rather vague, to put it mildly, assumptions. That such a strange thing could happen can be explained only by the fact that in the two-hundred odd years between Newton's *Principia* and Einstein's relativity theory classical mechanics had invariably proved its validity in practice with such accuracy and built up into such a stupendous, well-proportioned edifice that physicists looked upon the slightest suggestion of some fallacy in its foundations — the Newtonian laws — as on a preposterous and evil heresy.

As a result, worship took the place of scientific analysis: "God said 'Let Newton be', and all was light." It might be claimed in justification that, unlike mystical worship, worship of Newton found constant confirmation in the real facts of life. Be that as it may, but the truth remains that some of the fundamental propositions of mechanics were formulated by Newton rather ambiguously. Mathematicians would never have tolerated ambiguity in the fundamentals of their science, but physicists preferred to pretend that there was nothing amiss.

This, of course, was not because physicists were more "thick-skulled" than mathematicians. Simply the very quality of a mathematician's reasoning causes him to seek strict logical consistency; a physicist is quite satisfied if his theory provides an adequate description of real phenomena and he usually doesn't care very much about precise definitions of "self-evident" things. The concepts of length and time, for instance, were as clear as twice two is four. Yet Einstein has shown that these "simple" and "obvious" concepts are far from clear.

Two questions arise from what has been said:

First, how could Newton's laws operate at all in practice if, as we claim, they were ambiguously formulated?

Secondly, are we really to believe that Newton — the great Newton! — was as naive as is implied? Aren't we distorting the truth?

The answers to these questions will be forthcoming if we recall Newton's method. His aim was first and foremost to establish the principle, to deduce the general law from experimental data. The principle was needed in order to apply it to new investigations of natural phenomena. As a physicist he disliked speculative reasoning of a general nature. He was primarily interested in the practical application of a law. This may explain why Newton did not care very much about providing logically water-tight definitions of the basic concepts.

The important thing was to formulate the laws with sufficient clarity to be able to operate with them. Let there be some faults in the formulation of the principles of motion. Newton was not worried by the thought that the concept of mass actually remained undefined and that nothing was said about the concept of length: the layman, to say nothing of the physicist, knew exactly what he meant.

He defines the concept of force so carelessly that he fails to observe that he has merely paraphrased his own first law: "An impressed force is an action exerted upon a body, in order to change its state, either of rest, or of uniform motion in a right [straight] line."

He simply had no time to go into the details. His task was to create the science of mechanics and solve concrete problems.

One gains the impression that he was in a hurry to do the weary job of defining the basic physical quantities in order to get down to real work. As to the system of axioms in mechanics — let the future generations supplement it.

Here the author expresses his personal opinion, hence this passage should be treated with care.

Of one thing he was quite sure: his laws provided a basis for studying and describing every known type of motion, they were clear enough for that.

People may argue about the accuracy of the basic propositions of mechanics. Some might say, like scientists up till the end of the nineteenth century, that Newton's system is the best that man could ever devise. Or, as we shall see later on, it can be subjected to shattering criticism. But the fact remains that for more than two centuries not a single experiment conducted by any physicist offered the slightest challenge to the validity of Newton's laws. And whatever we might say further on, we should always remember that "God said 'Let Newton be', and all was light".

Before proceeding further, I should like to forewarn the unwary reader. The next few pages are perhaps the most difficult in the whole book. They may seem dull and boring. Unfortunately, they are essential for an understanding of all that will follow and, most important, for an understanding of Einstein's ideas.

Let us first review Newton's basic laws, or axioms, of motion.

1. *Every body continues in its state of rest, or of uniform rectilinear motion, unless it is*

compelled to change that state by forces impressed upon it.

2. The alteration of motion is proportional to the motive force impressed; and is made in the direction of the right line in which that force is impressed.

3. To every action there is always opposed an equal reaction; or, the mutual actions of two bodies upon each other are always equal and directed to contrary parts.

This is how they were formulated (in Latin) by Newton; this is essentially how we were taught them at school, though probably in different words. However, as long as the basic physical concepts employed in formulating them have not been defined, the laws themselves are as intelligible to us as the mysterious inscriptions on the Maya temples in Mexico.

Newton realised this, of course, and his laws of motion are accordingly preceded by a set of definitions of basic concepts. But, as stated before, Newton's nomenclature is not perfect, insofar as some things are redundant, others are absent, yet others are wrong or meaningless. Unfortunately, we have so much work ahead that we have time only for a cursory review of Newton's approach, without going into the details.

Drawing up a syllabus, which will cover three chapters.

And so, what basic concepts employed in the axioms of motion must be defined and analysed?

First of all, the concepts of length and time, then motion, force, and finally, mass. Furthermore, we must define velocity and acceleration. Then we will analyse the laws of motion and attempt to establish their physical meaning with greater accuracy.

Before proceeding with our programme, one last remark is necessary. Our task is not to provide general or ideal definitions. Far from it. All we want is to comprehend the physical meaning of Newton's principles and, as far as possible, the physical content hidden beneath the symbols and concepts we will be employing.

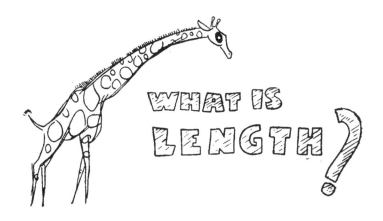

We shall begin with length (or distance).

Newton gives no answer to the question, "What is length?" This is unfortunate, for the question is prompted not by the idle speculations of a dithering scholastic. It is a very real physical problem. We shall take a purely pragmatic approach to it. What we want to know is how to measure the distance between two points, that is, the length of a physical body.

Luckily the determination of length concerns geometrists as well as physicists, therefore there exists a precise mathematical definition (mathematicians hate ambiguity).

Definition. The length of a line segment is a number which is attributed to the segment *in the process of measuring.*

The first surprise: the definition of length.

The recipe for measuring a line AB is:

1. Select a measuring rod M (say, one metre);

2. Divide the measuring rod into n equal parts (into 10 decimetres, for example); we can denote them $\dfrac{M}{n}$;*

3. Lay off segments $AC_1 = C_1C_2 = \ldots = C_{m-1}C_m = \dfrac{M}{n}$ from point A on segment AB as far as possible and denote the number m of the last segment (say, $m = 18$);

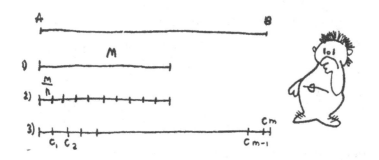

4. Increase n indefinitely (i.e., divide the measuring rod into centimetres, millimetres, etc.) and determine the corresponding number of units m (e.g., 183 cm, 1,834 mm, etc.).

The limit towards which the ratio

$$\frac{m}{n}\left(\frac{18}{10};\frac{183}{100};\frac{1,834}{1,000};\ldots\right)$$

This definition of length (or distance) is also valid in the special theory of relativity.

---

*Two segments are said to be equal if they can be superimposed through a process of motion. The properties of motion are in turn defined by the axioms of geometry. That it is possible to divide any segment into two equal portions, and consequently into any number of 2^n equal portions, is proved with the aid of other axioms of geometry.

tends is called the length of line *AB* as measured by the measuring rod *M*.[†]

The above is a typical example of the deductive method of reasoning, the principal method in mathematics. Some might find it long-winded and tedious. Others, however, will see in it the harmonious and exquisite beauty of mathematical logic.

In plain words, the definition of length consists in the following. Take a measuring rod the length of which, by definition, is unity. Apply it to the segment to be measured and see how many times it can be laid off. The number of times gives the length of the segment. To determine the exact length of the segment in terms of the measuring rod you must be able to lay off fractions of the latter, that is, you must know how to divide it into any number of very small and equal portions. That's all.

A rather important observation. A mathematical definition is translated into lay language.

That is how mathematicians approach the problem. For physicists, however, this precise definition is insufficient, and here is why.

Give me a measuring rod, you say, which I can apply to the line to be measured, and I'll tell you the latter's length. But what if physical conditions make it impossible to apply the measuring rod directly? How will you go about

---

[†]For length as here defined, the following important theorems can be proved:

Theorem No. 1. Every segment has length, which can be determined uniquely by any given measuring rod.

Theorem No. 2. The lengths of equal segments are equal.

Theorem No. 3. The length of a segment *AC*, which is the sum of segments *AB* and *BC*, is equal to the sum of the lengths of those segments.

Theorem No. 4. The length of a measuring rod is unity.

measuring the distance, say from the tower of Moscow's TV centre to the Kremlin without leaving your flat? Or what about standing on a railway embankment and measuring the length of a passing train? It will not wait while you try to apply your measuring rod to it!

Furthermore, ever present in the processes of length measurement is the concept of motion. Referring to geometry, you will be pleasantly surprised to learn that mathematicians consider motion to be a primary concept not requiring a definition. This, however, will not do in physics.

And one final remark. It is all very well for mathematicians: they operate with ideal geometrical lines. Their measuring rod neither expands with heat nor contracts under pressure: it possesses geometrical but no physical properties.

If we desire a precise definition of length acceptable to physicists we must take into account the real properties of the measuring rod and, therefore, postulate additional characteristics taking those properties into account.

After all that has been said it might seem that any attempt to define length and how to measure it is a fairly hopeless undertaking. But in science, as in life, we can put up with any difficulties on the road if we know whither it leads us. So far it is not at all clear whether physicists have anything to gain by engaging in such tasks as a thorough analysis of the concepts of length, time, etc. Maybe the whole problem is stuff and nonsense and of use to no one?

"Gentlemen," I hear some wizened nineteenth-century professor of physics lecturing us, "why not leave all these problems to the mathematicians? They hold the cards. Let them formulate precise definitions. We know what length is without worrying about definitions. It's plain as the nose on your face. Moreover, we know how to measure the length of a moving train without your directions and without applying a measuring rod. All we have to do is simultaneously mark the beginning and end of the train on the embankment. Then you can apply your measuring rod to your heart's content: the length will be that of the train. But in general, gentlemen, it's silly to reduce everything to the application of a measuring rod. Try and measure the distance between two mountain tops by your method. Nothing doing! You'll get nowhere without triangulation. And triangulation, you might like to know, involves the measurement of angles, which doesn't enter into your definition.

"We've got along without those definitions of yours since Newton's time and have been none the worse for it, no, siree! We've been measuring the distance to the stars and the length of microbes without applying your measuring rods to them.

"There's no denying, of course, that there is some sense in definitions. The idea of a measuring rod, for instance. We fully agree that

Some doubts. Incidentally the author becomes facetious.

a standard of length is necessary. But for your information, sir, we know all about it. Moreover, we've been the keepers and guardians of the standard of length for a hundred years.

"This, however, is neither here nor there. You just stick to nature. Investigate the phenomena and keep clear of the fundamentals of mechanics. Better men than you have worked there. Newton, you might like to know!"

These sentiments may sound extremely naive, but so far we have nothing in hand to contradict them with. Experience, the entire experience of classical physics testifies against us. After all, people did get on without definitions.

Nevertheless, in this case the physicists proved to be in the wrong. It was Einstein who demonstrated that, until the theory of relativity was enunciated, the notions of the universe and space on which science was based were in fact all wrong.

Today no one doubts that the concepts of time and length require precise definitions and that there is no place for self-evident notions in physics. But only Einstein was able to drive this home, though generally speaking probably no scientist would think of denying it.

The nineteenth-century physicist was not interested in the basic fundamentals of his science primarily because he was convinced that no new principles would ever appear.

Incidentally, mathematicians showed themselves to be more principled in a similar situation. For two millenia geometricians strove to prove Euclid's fifth postulate (the postulate of parallel lines). Their considerations were more of an aesthetic nature than anything else. The parallel line postulate stood out among the other axioms of geometry because of its comparative non-self-evidence. This goaded the mathematicians. At least, there seems to be no other reason for their persistent attempts to prove the fifth postulate.

Moreover, the creators of non-Euclidean geometry (Lobachevsky, Bolyai, Gauss) arrived at their conclusions not because of any practical faults of Euclidean geometry but through purely speculative reasoning.

But whereas mathematicians were able to arrive logically at the idea of the possibility of different systems of axioms and different geometries for describing space, for physicists this was impossible. For one, the fundamentals of physics (its axioms) had still not been expounded. Secondly, the very nature of research work fostered a prejudice against scrupulously logical, purely speculative reasoning. Only the genius of Einstein showed physicists how to combine the two methods.

Now that we know that a detailed analysis of the basic propositions of classical physics is essential for an understanding of the theory of relativity, we can continue our discourse with a clear conscience.

Let us see how the mathematical definition of length can be further supplemented. We have operated with measuring rods and real physical properties, which change with temperature, pressure and other conditions. We may find that these properties are *always* changeable, even as a result of say, motion. Suppose, for instance, that we have two steel bars, one in Moscow and one in Leningrad. We

Some unusual, and therefore difficult, ideas.

bring the Leningrad bar to Moscow, compare the two and find that they are equal (that is, they coincide when superimposed). Then we carry the Leningrad bar by a roundabout route and find that it has grown shorter. A fantastic assumption, it might seem, but nevertheless not to be ruled out.

Or maybe it is not the distance but time that affects the bar. That is, the longer it is in motion, the shorter (or longer) it grows. Another fantastic assumption, isn't it? But if you stop to think you will realise that the assumptions seem fantastic only because we unconsciously rely on our experience in considering them. And our experience tells us that such things never happen.

Again remember that such problems cannot be disposed of by speculative reasoning. They can be answered only by analysing experimental data. And the sum total of the known facts of physics may be expressed in the following postulate.

A postulate which may at least serve to illustrate how subtle is the strictly axiomatic definition of length.

Postulate No. 1. A real physical bar may always be moved relative to a measuring rod along any given path and in such a way that when the motion ends the length of the bar is the same as it was before the motion began, assuming, of course, that all other physical conditions, such as temperature, remained the same throughout the motion.

In formulating this postulate we did not seek flawless precision. All we did was to more or less state the experimental fact: "If there are two equal bars in Moscow, one of them can be carried around the world with such care that on its return to Moscow the two will be found to have remained equal."

With our definition of length and this postulate we may claim that if bar A is identically equal to bar B, and bar B to bar C, then $A = C$; or, it is possible to compare the length of bodies at rest even when they are very far away from each other.

Worthy of note is the following aspect of the problem. We remarked previously that, unlike mathematicians, physicists have to deal with real objects and must therefore bear in mind their real physical properties. Our postulate declares, essentially, that the length of a physical rod taking part in at least some types of motion is constant, that is, it is the same after a displacement as before (which makes it suitable for a measuring rod).

Hence, with the definition of length supplemented by postulate No. 1 we can precisely measure and compare the length of any objects at rest relative to each other.[‡] So far, however,

---

[‡]To be absolutely consistent, we should note that there still remain a number of blanks in the definition of length. Here is but one, and a fairly important one.

You have probably observed that in defining length we introduced the concept of motion, for without motion it is impossible to determine whether two lines are equal or not or to apply a scale rule.

Geometricians take motion as a fundamental concept and define its properties by introducing several axioms (the axioms of motion). It is all very complicated as their "mathematical" motion is a rather incomprehensible transformation of mathematical space into itself. But we deal with the ordinary physical motion of real physical bodies and we prefer a simpler definition, even if it is not quite water-tight.

The problem, however, is not easily solved. If length is defined in terms of motion, it follows then that, for the definition to be flawless, motion should be defined without introducing the concept of length. I wouldn't say for sure, but it seems that it is impossible to give a strict definition of motion without introducing the concept of length.

the only method of determining length available to us is the physical application of a measuring rod to the object to be measured, and we still don't know how to measure the length of an object moving in respect to a measuring rod.

The first difficulty can be overcome immediately. To determine the length of an object at rest relative to a standard of length, we can make use of any method of mensuration which geometry allows: triangulation, for instance, without which accurate geodetical surveying is impossible. The idea of triangulation is simple and undoubtedly familiar to many of you.

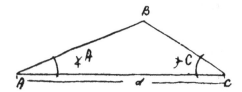

Say, you have to measure a line *AB*. For this, lay off *AC* — the datum line — at an arbitrary angle and measure its length by applying a measuring rod. Now measure angles *A* and *C*. With these quantities known you can determine

---

Consider: "The motion of a body relative to another body is defined as its change of position in respect to that body." There seems to be no other definition. But "change of position" cannot be analysed accurately without an understanding of length. Thus, in defining length, we introduce motion, while in defining motion we introduce length, and the circle is closed.

No mathematician would let such a fallacy stand. We, however; won't pay attention to such niceties even though we might feel vaguely disconcerted.

uniquely the parameters of triangle ABC, and from the formulas of trigonometry the length of AB.

Thus, if you have datum line AC — say, the distance between two Kremlin towers — you can determine with sufficient accuracy the distance to the spire of Moscow University (AB) without the tiresome, and often impossible, necessity of laying off a measuring rod. The convenience of the method of triangulation is obvious.

Of greater interest to us, however, is another consideration: the length of AB was, in fact, measured by a physical process totally unlike the process of measurement given in the definition. Instead of applying a measuring rod to AB we measured angles. Can we definitely claim *a priori* that the length AB obtained by triangulation will coincide with the length as measured by a measuring rod? Maybe the "triangulated" length of AB differs from its "measured" length? If we employ two different methods of measurement we cannot, of course, declare *a priori* that the lengths will coincide.

Recalling, however, that the theorems of geometry prove the equality of length, whether measured by triangulation or by applying a measuring rod, and remembering further that our geometry[§] is valid for our physical world, we conclude that the "measured" and "triangulated" lengths are the same.

In all practical problems, of course, our world is well described by Euclidean geometry.

Whether our geometry describes the surrounding world or not is decided by experience. If, in using Euclidean geometry for

---

[§]That is, the geometry whose equations we have employed in our calculations; for the trigonometric formulas of, say, Euclidean geometry and the geometry of Lobachevsky, are different.

our calculations, we found that the results of triangulation differed from the direct measurement of *AB* by measuring rod, we would be driven to the conclusion that the geometry of our world is non-Euclidean.

Everything said about triangulation holds good, of course, for any other method of determining length based on geometry.

Thus, knowing the geometry of our world (and reducing the problem of measuring length to it), we immediately obtain any number of methods for measuring length, for the theorems of geometry prove that, however length is measured, it is identical with the length obtained by the basic method of laying off a measuring rod.

To sum up. We have formulated a definition of length, borrowed from mathematics. It follows from the definition that in order to measure length we must have a physical standard of length, a measuring rod chosen by mutual agreement.

We have introduced a postulate which, though it may seem extremely vague, is necessary so as to rely on geometry in measuring the length of a body which is at rest relative to the measuring rod.

It was pointed out in passing that only experience gives us the geometry that describes our world.

It has been established that all recipes for measuring motionless bodies are reduced, thanks to geometry, to the basic method of laying off a measuring rod.

On the whole we seem to have learned nothing new. More, the suggested method of measurement cannot be used for measuring bodies moving relative to the observer. This latter situation may be formulated as follows: "It is impossible to apply a measuring rod to a moving object, as the object will simply move away."

True, it would seem possible to employ any of the many indirect methods of measurement, triangulation, for example. However, a careful examination shows that all methods of measuring the length of a moving body can be reduced to the following.

To measure the length of a segment moving relatively to an observer, it is necessary simultaneously to make intercepts of the initial and terminal points of the segment on an object at rest relative to the observer.

The distance obtained on the "fixed object" is measured by laying off a measuring rod. The resulting length is the length of the moving segment.

In practice this means: You are standing on a railway platform equipped with all possible measuring instruments. Passing by is a train whose length you have to measure. For this you:

(a) mark two points on the platform showing where the beginning and end of the train were at the same time;

(b) measure the distance between the two points and obtain the required length.

This is the definition of length of a moving body unconsciously accepted in classical physics.

Some people may not be convinced that all the methods employed by physicists for measuring the length of bodies moving in respect to an observer are reduced to the simultaneous fixing of the initial and terminal points of that body.

A very important definition. Length of moving bodies. The reader who gets as far as Chapter XII may find it useful to re-read this part.

If you belong to this category, you will have to take this for granted.

Now note the following. If we agree that, as a matter of principle, the discussed method is the only way of measuring the length of a body moving relative to an observer, then two questions immediately arise:

first, we must define what we mean by the simultaneousness of two or more events and how time is measured at all;

secondly, where is the guarantee that the new instructions for measuring will give the same result as the tried method of applying a measuring rod?

Before answering these questions attention must be drawn to an important consideration. Failure to understand it often has led to a rejection of Einstein's theory.

As the old definition of length (laying off a measuring rod) is unsuitable for moving bodies, we were forced to give a new definition for "the length of a segment moving relative to a measuring rod", introducing thereby a new physical concept. This concept arises from the new mode of measurement. It is important to realise that we cannot, we have no right to, assume that the length of a moving body must necessarily coincide with the length of a motionless body as defined before. Only experience can tell us whether these lengths will coincide or not.¶

¶Running ahead, we could note that before Einstein it was taken for granted that length was an absolute (*a priori*) concept, although this was supported only by unconscious reference to everyday experience.

After Einstein it became clear that length is a relative concept. The length of the same line segment may be different depending on the frame of reference in which

The definition of the length of a moving body is such that no amount of logical reasoning can prove that length to be in fact the same as the length of a fixed body. In effect, these are two different physical concepts.

In classical physics (the physics of small velocities) the length of a moving body is defined (or rather, unconsciously defined) in such a manner that it coincides with the length of the body at rest. Experience shows this to be true. But when velocities close to that of light were studied the picture changed. Lengths were found to vary. This was very strange, but hardly more...

In order to understand what is meant by "length of a moving body" we have to establish:

1. What is time?

2. What is simultaneity?

Standards. Their importance. Some history.

First, however, it would be useful to give some attention to standards. Most of you probably know that every self-respecting country has a National Bureau of Weights and Measures or some equivalent office where there are kept most carefully the standards of length, weight, time and other physical quantities.

Not all of you, though, may have asked yourself whether these bureaus serve any practical purpose or they merely represent a triumph for pure science.

---

it is measured. And this is confirmed by experience. In classical physics length was assumed to be an absolute quantity simply because at velocities much less than the velocity of light the length of a body moving relative to a measuring rod is almost exactly the same as that of a body at rest relative to that rod and it is impossible to detect the difference.

You may have heard people saying that a Bureau of Weights and Measures should, of course, have an accurate clock, but that the standard metre could well be abandoned. No one uses it anyhow.

Such remarks are graphic illustration of our tendency consciously or otherwise to generalise our personal experience and formulate rules from the customary, conventional facts of life (in this case the habit of checking time against a radio time signal).

The practical importance of the standard of length (as, of course, the standards of other physical quantities) is fairly obvious, however, so there is no need to waste our time refuting such ideas.

The metre as a standard of length — together with whole metric system of measurement — was introduced during the Great French Revolution. It was defined as one ten-millionth (1/10,000,000th) of a meridian quadrant of the Earth. In other words, the fundamental real object taken as the standard of length was the Earth: a rather ponderous measuring rod. There is an interesting anecdote in connection with this strange choice of a standard of length.

The president of the commission which worked out the metric system and the author of the system was the celebrated French mathematician Laplace. For his researches he needed accurate measurements of the Earth's meridian. But France was engaged in costly wars and no one, of course, would agree to finance an expedition with so nebulous an objective.

The solution to such a problem of immediate practical importance as the introduction of a new system of measures, however, was an entirely different matter. Until the metric system was introduced havoc reigned in France and in the world, and units of measure, all of them equally inconvenient, varied from country to country and almost from town to town.

Laplace (who, incidentally, was an able politician though, mildly speaking, neither an overscrupulous nor a high-principled statesman) easily convinced whomsoever it concerned of the need to measure the Earth's meridian for the purpose of introducing a new system of measurement.

The meridian (or rather an arc of the meridian) was measured. Science obtained some very important data, and mankind, a very convenient system of measurement, which spread rapidly throughout the whole of Continental Europe.

The English, who respect custom and tradition, have stuck doggedly to their inches, feet and yards (though Lord Kelvin once remarked caustically that if there was anything more nonsensical than the English system of measurement, it was the English monetary system).

A discourse on standards of length. Deliberations.

When the meridian was duly measured, a platinum-irridium bar was manufactured and named "metre". This is our measuring rod, our standard of length. Representatives from many countries agreed to adopt the distance between two notches engraved on the bar as the unit of length. In the final analysis, every metric ruler and gauge in the world is a descendant of that metre.

Subsequent accurate measurements revealed that the metre standard is only approximately equal to one ten-millionth of a meridian quadrant, but it was decided not to alter the international prototype standard. Faithful copies of the standard metre are distributed among all the countries of the world. The Soviet Union is in possession of prototype metre No. 28, which is also a platinum-irridium alloy bar.

The most precise measuring instruments are calibrated directly from the national prototype. From these, in turn, less precise instruments are calibrated, and so on down to the simplest wooden school ruler at the foot of the hierarchic structure on whose summit resides the International Prototype Metre in Paris.

And what will happen if the International Prototype is destroyed or, simpler still, changes its properties? Nothing of consequence, as there are many national prototypes. A minor "coup d'etat" would result, and by general agreement the British or, say, Soviet copy would be proclaimed the International Prototype Metre. As this new standard metre coincides with the one now in Paris to an extremely high degree of accuracy, the "change of government" (i.e., of the standard) would not affect the "people" (the measuring instruments) and everything would continue as usual. In the very unlikely event of the standard metre and all the copies going wrong, the standard could be restored by remeasuring the Earth's meridian. This, however, belongs rather to the realm of fantasy.

But if, say, the Moscow metre expanded slightly owing to a change in temperature (something less inconceivable) there would be considerable confusion in the Soviet Union until the error was traced to the prototype.

That is why the standard metre bars are kept in underground vaults where the temperature is maintained as constant as possible and tremors and other disturbances are excluded. (That, incidentally, is why metre bars are made of alloys possessing minimum thermal expansion.) In short, standards of measure are kept in conditions designed to guarantee the stability of their properties.

They say that the room in Paris where the standards of physical quantities are kept is locked with three locks. The keys to these locks are in the possession of three officers of the International Bureau of Weights and Measures. The room can be entered only in the presence of all three. This, probably, is one of the rare cases when bureaucratic methods are welcome.

Everything said about the standard of length refers equally to the standards of other physical quantities. The approach is the same. Physicists take some physical object or process whose properties are preserved constant and declare: "This is the unit of length (or mass, electrical resistance, time, etc.). We hereby decree this and henceforth all measurements are to be conducted according to the unit." It goes without saying that if the properties of a standard of measurement were allowed to change there would be such confusion that the standard keeper would be run out of town together with his standard.

To conclude our story of the standard of length, you would probably like to know that in October 1960 the General Conference on Weights and Measures adopted a new standard value: the wavelength of an orange line in the spectrum of krypton-86. This value is used to define the metre as being 1,650,763 of such wavelengths. This is a very convenient standard as physically it is constant for any place on Earth and, furthermore, it is always available. Its main shortcoming is its infinitesimal size (10^{-4} cm).

A few words about the history of systems of measurement.

A historian reads in an ancient Greek manuscript: "The lighthouse of Alexandria was 0.8 stadium high." Even without the previous discourse you will realise that this announcement carries very little information. How much is a stadium anyhow?

You may find no end of references to the stadium as a unit of length, you may learn that a stadium consisted of so many cubits. But as long as you are not shown a real object and told that, for example, "this sword is one cubit long and a stadium consists of 360 cubits", you will be as ignorant as ever.

Attention is drawn to the number 39,681 as an example of impotent accuracy.

Furthermore, you must be sure that the standard has not changed. But even this is not sufficient. For instance, you read in another book that at its narrowest the Hellespont was 4 stadia wide. The Hellespont — the Dardanelles of today — is there to measure it and it is hardly likely that the width of the strait has changed much since the ancient Greeks. But how are we to know the accuracy with which they measured it? This example brings to mind an amusing story about the determination of the length of the Earth's meridian. At school I was taught that Eratosthenes of Cyrene determined the circumference of the globe with amazing accuracy. He found it to be 39,681 kilometres.

One thing, however, is not so clear: how do we know this? Eratosthenes himself gives us the length of the meridian as 252,000 stadia. Evidently someone has found out how much one stadium is. Referring to an encyclopaedia we find a very "precise" definition of a stadium: Stadium, we read, an ancient Greek measure of length equal to 174–230 metres!

Rather taken aback by this degree of accuracy, I hastened to an expert on ancient Greece. He thumbed through several volumes before answering my query. It seems that we know hardly anything of the system of measures used by the Greeks. We are not sure that there was a uniform system in the Hellenic world. Did the Greeks have a standard of length? How long was a stadium? And, finally, with what accuracy did Eratosthenes compute the Earth's meridian? True, we find a standard cubit hewn on the walls of several Egyptian sarcophagi, but no one seems to know whether it was used in Greece or not. In general, historians are not very interested in the history of weights and measures, probably regarding them as of secondary value. This is unfortunate for the state of the weights and measures of a nation is at least an indication of the development of trade. Extensive trade, especially money trade, is impossible without some standards of length and weight.

We know some interesting facts about the standards of length used by the Arabs. They had as a standard the thickness of a hair from the muzzle of an ass. We hardly need to prove the fallacy of such a standard, which depended wholly on the individual qualities of this or that ass which, experience tells us, differ greatly.

Even stranger from our point of view was the standard accepted by the ancient Mongols: a day's journey on horseback. There was no uniformity there, of course, even though it may have suited Genghiz Khan's warriors.

Another interesting feature is the almost universal practice of defining units of length by reference to various parts of the human body. The yard, they say, was the distance from the tip of the nose of King Henry I to the tip of the forefinger of his outstretched hand. The name of the foot is self-evident, cubit comes from the Latin "cubitus", which means "elbow". The fathom was originally the distance between the finger-tips of the two hands when the arms are outstretched in a straight line. Laplace's metre is, of course, also of the same order as the height of a person.

And so much for length.

Probably no one before Einstein tried to analyse what was actually meant by time. Physicists were satisfied with Newton's vague definition. The first comprehensive criticism of the basic propositions of Newtonian mechanics was launched by Ernst Mach, who is also known as the author of a reactionary idealistic philosophical system bearing his name. His philosophical system and the general physical conclusions deriving from it (such as his denial of the existence of atoms) are not, mildly speaking to his credit.

Now we shall proceed to discuss the concept of time in physics. This is a problem of primary importance. The reader will probably find many surprises and, consequently, difficulties.

His critique of Newton, however, was on the whole undoubtedly progressive, and on several occasions Einstein mentions that he was greatly influenced by Mach.

The most important portions of Mach's work are those where his attitude is negative and in which he points out the fallacy and meaninglessness of Newton's propositions. He failed to bring clarity into the basic propositions of mechanics, for his own assertions and claims were often erroneous or meaningless. His merit is that he made the first breach in the wall of blind worship in Newton.

Here is what Newton has to say about time: "Absolute, true, and mathematical time, of itself, and from its own nature, flows equably without relation to anything external, and by another name is called duration: relative, apparent, and common time, is some sensible and external (whether accurate or unequable) measure of duration by the means of motion, which is commonly used instead of true time; such as an hour, a day, a month, a year."

Newton on time. Absolute time as an example of a meaningless definition.

And another remark of interest to us: "The natural days are truly unequal, though they are commonly considered as equal, and used for a measure of time."

The definition of absolute time is an excellent example of how Newton the philosopher contradicts Newton the physicist.

The physicist recognises only those physical concepts which lend themselves to actual investigation. The philosopher imposes on the physicist a completely meaningless concept of absolute time, the very definition of which makes it impossible to state anything positive about it. The same goes for his definition of space.

The physicist holds his peace and finds consolation in the fact that he never really applies these concepts to the solution of real problems. True, sometimes the physicist ventures to make some sceptical remarks contradicting the philosopher's views, but the latter very quickly calls him to order.

It could be noted also that philosophically Newton's concept of absolute time is unsuccessful because, if we are to believe him, time is in no way associated with matter.

With our experience gained in formulating the notion of length we can deal fairly quickly with time. Our mode of action is analogous in both cases.

First of all, we need a standard of time, something like our measuring rod.

We are dealing here with the "physical" not the "philosophical" definition of time.

For a standard of time, we must take some physical process (e.g., the Earth's rotation about its axis) and declare: "The duration of this process is the unit of time."

Thus we obtain a clock standard. Like the length standard, we must preserve its properties unchanged. In other words, in order to make such a clock we must base it on a physical process which can be repeated identically and indefinitely. It doesn't matter whether this process recurs naturally (as the rotation of the Earth) or due to artificial conditions (as the swaying of a pendulum).

You will have observed that the definition of the standard of time is very like that of the standard of length. Now that we have a time standard we must provide a recipe for measuring time. But first recall Newton's observation: "The natural days are *truly* unequal, though they are commonly considered as equal."

After what has been said before this observation seems most inappropriate, for if we have taken the natural day for a time standard we have thereby assumed that the days are equal *by definition*. It thus seems useless to ask whether the days are *truly* equal or not.

We begin a scrupulous analysis of an apparently obvious problem and, naturally, complicate matters.

But let us not jump to conclusions. It has been mentioned on several occasions that the main requirement imposed on a standard unit is stability of its properties. And so it is high time to answer a question which may be on the tip of your tongue by now: How are we to establish whether the properties of some standard of measurement (say the length standard) have changed or not? What, for that matter, do we mean by properties changed or unchanged? Changed in relation to what?

An unexpected but pertinent question is raised.

The standard of measurement defines the physical quantity to which the unit of measurement is referred and it is our duty to trust it above all. By establishing a physical object to which a unit of measurement is referred we thereby put an end to any discussion. The metre standard in Paris, we can now say, is *by definition* the unit of length even if it expands from heating.

Unfortunately, it often happens that a logically sound proposition may run counter to the facts of life. The most vivid examples of this are found in mathematics. We could concoct no end of logically flawless geometries, but there is only one geometry in the real world.

Therefore if the standard metre suddenly ceased to coincide with all its copies, which, however, would continue to be equal to one another, the physicist would declare the standard to be at fault and chose a new one.

There is only one way in which a change in the standard can be detected: by comparing it with some object or objects whose properties are known for sure to have remained unchanged. Any observed discrepancy will mean that the standard has changed. This can be illustrated by the following example.

Imagine a dozen or so three-year-olds who have chosen the height of one of them as their standard of measurement. From their point of view it may turn out that the height of any member of the group remains fairly constant. Moreover, if their "standard" happens to outpace the others in growth they will find to their dismay that they are "growing downwards". Soon, however, they will notice that the things about them — chairs, tables, parents, the room, a dog — seem to become smaller (or rather not so large). The cleverest will conclude, and maybe convince the others, that actually they are all growing. They will then discard the old "standard" and choose a new one, say the mark made by someone's dad on the door jamb. They will regard the distance from the floor to the mark as absolutely constant since its relative length, as compared with other inanimate objects, remains the same.

Scientists reason along approximately the same lines. The physicist has at his disposal hundreds of objects whose relative dimensions are known to him. Referring again to length, there is the International Prototype Metre, scores of copies and hundreds of thousands

of objects whose length has been measured by reference to the standard. For example, the Earth's meridian is approximately equal to 40,000,000 unit standards of length. The number will change only if the Earth or the metre standard change.

The relation of the standard of length and the lengths of many objects with different properties (wavelengths of the spectrum, the Earth's meridian, metre prototypes, etc.) have been established by direct measurement. If the relative dimensions of these objects remain unchanged, then we can claim that their respective lengths have not changed. For, if the relative dimensions have not changed, any change in length must be proportional in all the objects under consideration. And we have no reason to suspect that there exists some unknown cause capable of proportionally altering the length of the most diverse natural objects. It is appropriate to recall Newton's words: "Occult qualities have no place in experimental philosophy."

On the other hand, if the relation between one of the prototype standards and all the others changes, we say that the former's properties have changed. For the sake of convenience one object is taken as the "main" standard of measurement and all others are referred to as its copies.

If the relation between the "main" standard and its copies has changed, we say that the properties of the former have changed and we will act as the children in our story did and choose one of the copies as the standard of length.

All this, of course, refers not only to length but to time and all fundamental physical quantities as well.

Actually, every physicist is in possession not of a single standard of length, time or any other quantity but of a whole family of standards. An important feature of such a family is that physically its members should have as little in common as possible. As long as peace reigns in the "family", we say that the properties of each member have not changed. When some new physical object is examined and the relation between it and every member of a family of standards is found to be constant, the object can be accepted as one of the family. But as soon as any "member" makes a "breach of the peace", it is immediately ostracised.

Hence, we can say: "The length of an object can be said to have truly changed if its relation to the whole family of standards of length has changed."

Similarly, having established a discrepancy in the relation of a "natural day" to the whole family of standards of time (a sidereal day, the half-life of radioactive elements, the oscillation period of a piezoelectric crystal, etc.), we say that "the natural days are truly unequal".

The meaning we place in these words, however, is not the same as Newton did. He spoke of the changing duration of the natural days in respect to some mysterious "absolute time". We refer the changing duration to the family of standards of time. This, and only this, is what is meant by the words "truly unequal".

Note also the following. Having analysed the concept of a family of standards, we no longer judge the properties of a standard by our subjective sensations. Now we rely on objective properties of the physical world. That the relative dimensions of standards of measurement do not change is a real fact of the physical world. Hence, our claim that the properties of a family of standards do not alter is an objective truth.

Analysing a hypothetical case that could have been advanced by an agnostic.

At this juncture we could pose a rather strange and aggravating question.

Imagine that inhabitants of the Andromeda Nebula (whose civilisation, if we are to believe sci-fic writers, has evolved far ahead of ours) possess fantastically powerful telescopes. Imagine further that they have established that *from their point of view* the portion of the universe familiar to us is continuously pulsating. Using their standards of measurement, they find that all our objects and the distances between them expand and contract perfectly uniformly. They have found that, "as they see it", all linear dimensions, including, of course, the linear dimensions of our family of length standards, expand and contract uniformly by a factor of 100.

To the denizens of Andromeda this will be as objective and real a fact as to us is the assertion that the linear dimensions of a whole family of standards are constant. For, if there is agreement within such a family, we have no way of establishing the fact that the linear dimensions are regularly changing.

What would *actually* be taking place in our hypothetical case? Not from our point of view or as viewed by the Andromedans, but in the real, objective world which does not depend on anyone's point of view?

Maybe we could find an answer if we recalled that many physical laws are associated with distance and other physical properties, such as the laws of interaction of particles, the gravitational law and Coulomb's law. Since from *our point of view* all physical laws and properties remain constant, the Andromedans would additionally observe that the mass of bodies, the gravitational law, the laws of nuclear interaction — in short all the physical laws and properties of our universe — are subject to regular alterations. They would come to the conclusion that our portion of the universe is an extremely strange world with most peculiar laws.

And all the time we would continue to assert that the laws of nature do not change. Is this possible? It is a very peculiar assumption, but then, who knows?

Scientists from two different worlds meeting on neutral ground would observe something else (say, the pulsation of both worlds). What would actually be taking place in the real world which is independent of any observers?

Actually and independently of observers the following would be taking place: all physical relationships between the parts of the universe would be changing in some specific way. Every observer would describe those changes according to the notions and laws formulated by him in the exploration of his part of the universe.

All observers would thus be proceeding from objective facts, from their own purely objective definitions of fundamental physical notions. Each one would describe what was actually happening in his own language, for physical laws and concepts constitute our language for describing the real world.

Thus, in our hypothetical case the Andromedans would be as right in their judgement as we earthmen. Why are we inclined to think that the case considered is purely hypothetical? For the sole reason that nothing in our experience gives cause to assume otherwise.

After all that has been said, even the sidereal day is "compromised". The most reliable standards are quartz, molecular and atomic clocks.

Hence, we claim confidently that the sidereal day is *truly* constant. But we remember the meaning of the word, and we shall not be very much astonished if we learn that *from the point of view of an observer in other physical conditions this is not so*. We will readily come to terms with such an observer concerning the point of view from which to describe the real world. Most probably the simplest and logically more consistent system of laws will be accepted. This, though, is secondary. The important thing is that, whatever description is accepted, it should reflect reality.

You see that an explication of the seemingly clear word "truly" has required a detailed analysis.

Was it worth wasting so much time? Yes, because we have been able to gain a clearer insight into the nature of physical concepts. Now you will not be discouraged when we discuss the ideas and conclusions of Einstein. All this has been done for future benefit.

Let us now formulate a recipe for measuring time and be done with it. To measure length we had to be able to divide a unit standard into any number of equal parts. Similarly, to measure time we have to learn to divide the standard of time into equal small portions.

In the case of length we were assisted by geometry, but time is a concept that doesn't exist in geometry, so we shall have to get along without the help of mathematicians.

To divide a standard of any physical quantity into equal portions means, in effect, to introduce a smaller unit into the family of standards. We can always find such a unit amidst the countless physical processes going on around us.**

If we have at our disposal a standard for measuring time (a clock, in short), then in order to measure the duration of any physical process we must observe the readings of the clock simultaneously with the beginning and the end of the process. The time interval by the clock gives the duration of the phenomenon.

But what is meant by two physical events taking place *simultaneously* at some point of space? It would seem fairly obvious, but to assure the reader that his trials are not in vain, note that this "obvious" concept is the focal point of Einstein's theory.

Consider the following definition.

Definition 1. *Two events such that, generally speaking, either one could be the cause or the*

Getting back to a recipe for measuring time. This passage is very important for an understanding of Einstein's theory.

This definition is valid in Einstein's theory.

**We know from experience, for instance, that the rotation of the Earth with respect to the fixed stars through any very small angle is a process belonging to the family of time standards.

effect of the other, which take place at one point in space can be simultaneous only if neither is the cause nor the effect of the other.

Clear and logical, isn't it? Now we shall have no difficulty in comparing the time on two clocks located at the same point of space. But what if our clocks are at different points? Quite simple it would seem: take simultaneous readings of both clocks. How is this to be done? We have just defined the concept of simultaneity of two events taking place *at one point.* And what is meant by two events taking place simultaneously *at different points* in space?

We need another definition.

But this definition had to be altered substantially.

Definition 2. *Two events taking place at different points in space can be simultaneous only if neither is the cause nor the effect of the other.*

We have a definition, but one vague point remains. Let an event A have taken place at some point X_1 in space. Generally speaking, some time will have to pass before this becomes known at some other point X_2.

This rather abstract reasoning can be illustrated by a simple example. Not long ago I read a short newspaper item entitled "Stop Watches Click Simultaneously". Before, it said, the judge at the finishing end of a race track was unable to fix accurately the time of the shot of the start pistol. Decimals of a second would pass before the sound reached him (0.2 sec for a 110-yard track). Now a flashbulb is synchronised with the pistol and the judge at the finish tape sets his watch going as soon as he sees the flash.†† The two events (the shot and

††Strictly speaking, it takes several fractions of a second for the flash to reach the judge's eye. Furthermore, human

the starting of the watch) are assumed to take place simultaneously. If we were very exacting, however, we would have to recognise that the shot at the start (event A at point X_1) and the starting of the stop watch (event B at point X_2) are still not simultaneous. For one, it took time for the light to reach the finish: a very short time, indeed, but nevertheless appreciable. Furthermore, the stop watch was set in motion by the flash at the start (there is a relation of cause and effect between event A and event B).

The most surprising conclusions, however, arise if we assume that there exists a maximum velocity at which signals can be transferred (the velocity of light?). Then there must exist some minimum time of information t_{inf} in which the signal travels from X_1 to X_2. But if that is so, then any two events at points X_1 and X_2 separated by a time interval less than the time of information cannot be connected by relations of cause and effect. (It is impossible to notify the judge at the finish tape about the start faster than it takes light to travel the distance. And while the beam is en route....)

Thus, if we stick to our definition, to event A at point X_1 there may correspond an indefinite number of simultaneous events at point X_2. Hence, our definition is not unique.

You see that we have to discuss such an "obvious" concept as simultaneity. Strange as it may seem, but the simultaneousness of two physical events taking place at two points in space is far from self-evident.

This assumption lies at the root of the theory of relativity.

---

reaction is also rather low, but we assume that our judge is an ideal automaton.

This hypothesis is wrong.

The proof is wrong. Newton's gravitational laws are only approximately correct.

But such a painstaking analysis is much too tiresome. Let us introduce the following hypothesis: *There may in principle exist signals which propagate with infinite velocity.*

Now we can define uniquely two simultaneous events at different points in space, whence we can compare two clocks at different points and determine whether they show the same time.

How did we arrive at our hypothesis? Strictly speaking, physicists had practically no proof to support it, with one exception. We shall see further on that the gravitational theory rests completely on the proposition that gravity propagates with infinite speed, and Newton's gravitational theory is absolutely water-tight. This has been confirmed by experience. So gravity is an example of a signal possessing infinite velocity, whence the postulate of the propagation of a signal with infinite velocity is also confirmed by experience.

With this we shall wind up our definition of time and simultaneity and draw a few conclusions.

We have no reason to assume that our system of concepts of length and time is valid in itself. All its propositions and postulates stem from experience and nothing but experience. We may therefore be sure that they are true, at least approximately. If, however, new experiments reveal that, strictly speaking, our conclusions are wrong, then we will change them. There is nothing terrible in this. We will merely note once again that our postulates describe the real world only approximately and will be gratified by the fact that this is demonstrated graphically by "experiment, that supreme judge of physics."‡‡

---

‡‡We have declared repeatedly that all postulates of classical physics, which have proven to be erroneous from the point of view of the theory of relativity, stem from experience. On the other hand, Einstein's theory also stems from experience.

Is this possible?

The thing is that, when small velocities as compared with the velocity of light are involved, classical physics describes the world to a high degree of accuracy.

Our postulates and definitions are based on such velocities. That is why the fallacy of the postulates remained hidden for so long.

CHAPTER IV,

the shortcomings of which are atoned for by the epigraph. It contains a rather dry and long-winded explanation of what is meant by a frame of reference; the important idea is brought home that without a frame of reference any talk of mechanical motion is absolutely pointless

NEWTON. MECHANICS
(analysis of basic concepts: motion)

There is no motion, a bearded sage
expounded.
Another held his peace, but back and forth
he paced.
It was a subtle argument, the greybeard
was disgraced:
No better answer could he have propounded.
'Tis an amusing anecdote, no doubt,
But in my mind another case does rise:
For daily do we see the Sun traverse the
skies,
Yet stubborn Galileo's truth we shout.

PUSHKIN

To begin with, some quotations. Sir Isaac Newton, *Mathematical Principles of Natural Philosophy* (General Propositions):

"Absolute space, in its own nature, without regard to anything external, remains always similar and immovable. Relative space is some movable dimension or measure of the absolute spaces, which our senses determine by its position to bodies, and which is commonly taken for immovable space....

"Place is a part of space which a body takes up, and is according to the space, either absolute or relative....

Newton's views on space and motion.

"Absolute motion is the translation of a body from one absolute place into another; and relative motion, the translation from one relative place into another....

"And so, instead of absolute places and motions, we use relative ones; and that without any inconvenience in common affairs; but in philosophical disquisitions, we ought to abstract from our senses.... For it may be that there is no body really at rest, to which the places and motions of others may be referred."

Thus speaks Newton the philosopher. It is he, not the physicist, who introduces absolute space and absolute motion.

Absolute space exists without relation to matter, without regard to anything external. It is an abstract, speculative concept, a mysterious receptacle of some divine principle. Newton the philosopher and the theologist believes in "the counsel and dominion of an intelligent and powerful Being" and in absolute space. He most surprisingly forgets his own rule: "We are to admit no more causes of natural things than such as are both true and sufficient to explain their appearances."

May the ghost of the great Newton forgive us the affront, but it is a fact that his definition of absolute space lacks physical meaning.

But now the phyicist speaks: "It is indeed a matter of great difficulty to discover, and effectually to distinguish, the true motions of particular bodies from the apparent; because

Newton on absolute and relative motion.

the parts of that immovable space, in which those motions are performed, do by no means come under the observation of our senses. Yet the thing is not altogether desperate."

The physicist, armed with his wonderful talent, now comes to the rescue of the philosopher. He shows how to define absolute space by finding a way of demonstrating absolute motion, i.e., motion in relation to absolute space. Here is his recipe.

Newton's famous experiment with a revolving bucket of water.

"The effects which distinguish absolute from relative motion are the forces of receding from the axis of circular motion. For there are no such forces in a circular motion purely relative, but in a true and absolute circular motion they are greater or less, according to the quantity of the motion. If a vessel, hung by a long cord, is so often turned about that the cord is strongly twisted, then filled with water, and held at rest together with the water; thereupon, by the sudden action of another force, it is whirled about the contrary way, and while the cord is untwisting itself, the vessel continues for some time in this motion; *the surface of the water will at first be plane, as before the vessel began to move*; but after that, the vessel, by gradually communicating its motion to the water [through friction], will make it begin sensibly to revolve, and recede little by little from the middle [from the axis of rotation], and ascend to the sides of the vessel, forming itself into a concave figure (as I have experienced)....

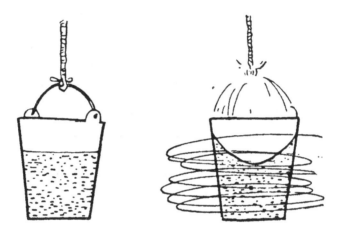

"At first, when the relative motion of the water in the vessel [with respect to the walls of the bucket — V.S.] was greatest, it produced no endeavour to recede from the axis; the water showed no tendency to the circumference, nor any ascent towards the sides of the vessel, but remained of a plane surface, and therefore its true circular motion had not yet begun. But afterwards, when the relative motion of the water had decreased the ascent thereof towards the sides of the vessel proved its endeavour to recede from the axis; and this endeavour showed the *real* circular motion of the water continually increasing, till it had acquired its greatest quantity, when the water rested *relatively* in the vessel."

We now have a criterion of absolute motion — *centrifugal force*. The existence of a centrifugal force can always be determined by the shape of a moving body or the internal stresses and strains that appear in it.

Maybe Newton has actually found a way of defining absolute motion, and hence, absolute space? Maybe all that is wrong with his definition of absolute space is its unhappy wording and he does indicate a real way of defining absolute space and motion?

Experience tells us that centrifugal forces appear in a body when it rotates relative to the fixed stars. Maybe it is reasonable to speak of absolute motion when it is relative to the stars? Maybe the fixed stars define absolute space?

Attention! An important, cardinal question.

To sum up, *does there exist a type of motion which we could call absolute or is all mechanical motion relative?*

Without anticipating the answer, let us analyse the "apparently well-known" concept of motion.

To begin with, a description of "mechanical motion" makes sense only if it is referred to some real physical bodies which can be assumed to be fixed. This is what is known as a "frame of reference". As long as a frame of reference (or "fixed" body) has not been indicated, the words "a moving body" are meaningless.

Attention again!

It is obvious from what has been said before that Newton was well aware of this. In speaking of relative motion he actually introduced the concept of a frame of reference. More, he was the first physicist to realise the importance of reference frames. Galileo himself had no clear idea of it, hence, he had no clear idea of mechanical motion. He cut short his analysis just when the question "What is motion?" had to be answered.

The reason for this is that, despite its apparent self-evidence, the concept of a frame of reference is so abstract that it could appear only at a fairly high stage of scientific development.

Surprisingly enough, even in our days many people well versed in mechanics and capable of solving problems which would have baffled Newton are at a loss to say "What is actually happening": whether a locomotive is moving relative to the Earth or the Earth is moving relative to the locomotive.

So what determines our choice of a frame of reference? What real bodies can be assumed to be fixed?

The bodies that can be assumed to be fixed — the frame of reference — are chosen by us at our own discretion. In other words, the choice of a frame of reference is determined by our own convenience.

In analysing the flight of a projectile we choose a frame of reference rigidly connected with the Earth. In investigating the Earth's motion we attach our reference frame to the Sun. In studying the Sun we refer its motion to a frame connected with the stars.

Since the choice of a frame of reference is arbitrary, a train passenger and his friends standing on the station platform may with equal right claim to be at rest. The passenger may take a frame of reference fixed with respect to his coach, and in that frame the railway station (together with his friends) will be moving away from the point of origin of the frame. In a reference frame fixed with respect to the Earth, it will be the train, of course, that is moving.

If somebody claims that "actually" the train is moving, it is because intuitively, on the basis of our daily experience, we usually choose a frame of reference fixed with respect to the Earth.

The geocentric system of Ptolemy is an excellent example of the unreliability of such intuition.

This is to remind the reader that the question of an absolute reference frame and absolute motion is still pending.

But maybe among the countless possible frames of reference there is one (one!) unique and inimitable frame whose physical properties differ so greatly from those of the others that it may be considered absolute. *If there does exist such an absolute frame, then we may speak of true (absolute) motion and space.*

We have returned to the question raised before. Newton, you will recall, suggested a method for defining absolute motions (centrifugal forces!), but so far we are not in a position to judge if he was right. His definition of an absolute frame, however, does not suit us as it forces upon us the mysterious concept of absolute space. Therefore, refraining from solving this problem for the time being, let us give a most general definition of the process of motion.

A definition of motion which probably corresponds to Newton's "relative motion".

A given physical body is said to be in motion in respect to other physical bodies if it changes its position relative to those bodies.

Nothing new here, you see, just summing

up what has been said before. For complete satisfaction we must define precisely what is meant by the words "changes its position relative to other bodies". The answer is fairly simple.

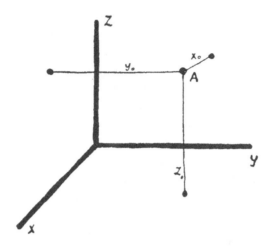

We rigidly attach a frame of reference to the body or bodies we assume to be motionless. Then we determine the coordinates of the body under consideration and define its "position". From our school geometry course we know about Cartesian systems of coordinate axes. In this system the position of a point in space is uniquely defined by its shortest distances to each of three mutually perpendicular planes. Other coordinate systems are often used in mathematics, but they all require three quantities, three coordinates to define the position of a point in space uniquely. Without going too deeply into mathematical or topographical explanations, we can say that a coordinate system is said to be "rigidly attached" to a body if the coordinate axes

This, by the way, is what we mean when we say that space has three dimensions.

are directed towards, or pass through, some constant specified point of that body. In order to attach a Cartesian coordinate system to the Earth, for example, we could place the point of origin at the centre of the globe, with the z axis pointing towards the North Pole, the x axis, toward the intersection of the Greenwich Meridian with the equator (0° lat. and 0° long.), and the y axis, towards the point 0° lat., 90° long. The origin, of course, could be placed in any other point of the globe and the axes pointed in any other direction and it would still be attached to the Earth.

Now we are in a position to give a more precise definition of motion.

A body is said to be in motion in respect to a given coordinate system if at least one of its coordinates changes with time.

The manner in which coordinates may change is described by that important characteristic of motion, velocity.

Without going too deep into details (which would require more mathematics than is generally permitted by the canons of popular literature), the concept of velocity can be introduced in the following manner.

We wish to determine the velocity of a body at some specified moment of time t_0. The procedure for this will be:

(i) Locate the body in some chosen frame of reference at time t_0, i.e., determine its coordinates.

(ii) Determine the body's position at some time t_1 (i.e., determine its coordinates at that time).

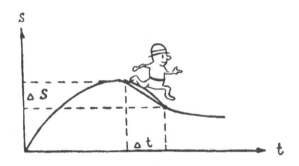

(iii) Determine the length of the straight line $\Delta S(t_1, t_0)$ connecting the two points.

(iv) Divide $\Delta S(t_1, t_0)$ by the time increment $\Delta t = t_1 - t_0$. The approximate magnitude of the velocity at time t_0 is $[v(t_0)] \approx \dfrac{\Delta S}{\Delta t}$ The smaller the time increment Δt, the more accurate the ratio $\dfrac{\Delta S}{\Delta t}$ which defines the velocity at time t_0.

In the limit, when $\Delta t \to 0$, the ratio gives the exact absolute speed of the body at time t_0. This is denoted

$$[v(t_0)] = \lim_{\Delta t \to 0} \frac{\Delta S}{\Delta t}.$$

The drawing illustrates the operations performed here for the special case of rectilinear motion. On the diagram the value of S begins to decrease at some moment. This means that the body has reversed its motion. The velocity at the highest point of the curve is zero. To the left of that point the velocity is positive, to the right it is negative. Note that if the approximate expression for velocity is used we do not obtain zero for the highest point of the curve.

It was mentioned before that magnitude alone is not sufficient to characterise velocity (the magnitude of velocity is what we generally call speed). We must also know the direction in which the body is moving from the initial point.

If motion is non-rectilinear, a body may follow a meandering path, and this means a change in velocity. A body is said to move with uniform velocity when both the speed and the direction of motion are constant (uniform rectilinear motion). Obviously, the direction of velocity is given by the direction of the segment ΔS.

And now we are ready for the most important part. The displacement $\Delta \vec{S}$, as we mentioned before, is described in a given frame of reference. The magnitude and the direction of $\Delta \vec{S}$ depend on the choice of the reference frame. That is to say, the displacement is a relative quantity which depends on the frame of reference in which it is described.

This should be familiar from the school course of physics, so we shall limit ourselves to a visual "railway" illustration.

The path travelled by a Moscow–Leningrad express in a frame of reference rigidly connected with the train itself is identically equal to zero (the train is always at the origin of the coordinate system, hence $\Delta \vec{S} = 0$).

Consider a frame of reference connected rigidly with a goods train that left Moscow at the same time as the express. The goods train, naturally, travelled slower, and when the express reached Leningrad it had only pulled in at Bologoye. The path travelled by the express will now be equal to the distance Bologoye–Leningrad ($\| \Delta \vec{S} \| = 325$ km).

In a frame of reference referred to the Earth the express will have travelled the distance from Moscow to Leningrad, i.e., $\| \Delta \vec{S} \| = 650$ km. But since the velocity is determined by the ratio $\dfrac{\Delta \vec{S}}{\Delta t}$, it therefore also depends on the frame of reference in which it is measured.

Incidentally, such examples may be more involved than clear-cut mathematical formulas.

And what about the time interval Δt? Does it also depend on the frame of reference? Is it possible, in measuring the time it took the express to reach Leningrad, to obtain one result for the reference frame referred to the Earth and another for the frame referred to the train? Or is the very question absurd? I hope that by now no one will venture to say so.

Time is a physical concept introduced on the basis of experience. In classical physics we assume that the time interval Δt is the same in all reference frames. This conclusion is based on the sum total of human experiences. But if, nevertheless, new experiments demonstrated that a time lapse Δt was of different magnitude in different reference frames, we might be surprised but not dismayed.*

In this connection we might recall a character from one of Mark Twain's books who was convinced that in the countryside time passed much slower than in the city. Complete ignorance of physics made it possible for him to enunciate his daring hypothesis. In any case, however, he too proceeded from his intuitive, if misconceived, experience with time.

In classical physics the concept of time is such that any time interval Δt is absolute regardless of the frame of reference.

It follows then that velocity, like displacement, is also a relative concept which changes from one reference frame to another. That is about all that we had to recall about velocity. Knowing the concept of velocity, we can by analogous reasoning define acceleration:

$$\vec{a} = \lim_{\Delta t \to 0} \frac{\Delta \vec{v}}{\Delta t}.$$

Repetitious statements. Conclusions and an unsolved problem.

Acceleration in respect to velocity is as velocity in respect to path.

Let us now see where we stand. We have stressed repeatedly the essentially trivial idea, which we shall nevertheless repeat once again:

---

*This was the reaction of physicists to the theory of relativity. But as long as velocities are much less than that of light, any time interval Δt may be said to be identical in all reference frames.

"Only when we declare some real physical bodies to be motionless, indicating a frame of reference, can we speak of mechanical motion. Without indicating a frame of reference the words 'rest' and 'motion' mean nothing at all."

You will have observed from the passages quoted from Newton's *Principia* that he himself appreciated the importance of a frame of reference. He assumed, however, that there existed some special, exclusive, wonderful and inimitable absolute frame of reference. He even suggested a method of measuring absolute (true) motions (the bucket experiment)!

We have not yet determined whether such a reference frame exists or not. It was the search for an answer to this simply posed question that led to the theory of relativity.

We shall see in the next chapter that the laws of mechanics are such that we are unable to proclaim some exclusive reference frame to be better than all the others.

There exist an infinite number of frames of reference, each of them, from the point of view of mechanics, as good as any other. They are called "inertial frames".

The question might then be asked whether it would not be possible to locate that mysterious absolute frame by investigating non-mechanical phenomena, say electrical, magnetic or gravitational.

Maybe there does exist a unique, preordained frame unlike any other? Possibly we could find such a frame if we investigated, say, electromagnetic phenomena?

From Chapter VII on we shall follow (unfortunately, very superfluously) the attempts to answer this question and the explorations which culminated in the enunciation of the relativity theory.

And so, once again we are confronted with the dilemma: "Is it possible by some physical experiment to locate such a wonderful, unique frame of reference whose physical properties would differ markedly from those of any other frame?"

CHAPTER V,

in which the author first dis-
courses and then professes amaze-
ment and calls upon the gentle
reader to follow suit

NEWTON. MECHANICS
(analysis of basic concepts: frame of reference)

> *Newton was fortunate, because the science of
> our world can be created only once, and it was
> Newton who created it.*
>
> ### LAGRANGE

Long-winded explanations are far from perfect
and the author is consequently beset with
doubts. Will not the foregoing elaborate and
rather dull analysis seem unnecessary? After
all, the contents of the whole of the previous
chapter can be reduced to several statements:

Any discussion of the mechanical motion of
a body makes sense only if a frame of reference
associated with some real body or bodies is
indicated.

More reiterations and the usual general remarks.

In the final analysis the choice of a frame of reference depends solely on its suitability for describing a given phenomenon.

If there exists some exclusive reference frame in which the laws of nature are represented with special lucidity (better than in any other frame), then such a frame of reference may reasonably be declared absolute, and we may speak of absolute motion. It remains to be seen if there is in fact such an absolute frame or not.

Furthermore, the main effort of our reasoning was aimed at elucidating the former proposition.

Can it be that we have merely been wasting time and energy to elucidate the "apparently well-known" concept of motion? That we have been forcing an open door only to lose our way in a maze of reservations, clarifications and explanations? Or is it this that comes from evil, as they say? Well, probably not.

Always remember that what seems self-evident may hide serious problems. The mathematicians were perhaps the first to realise this (recall Euclid's fifth postulate). In our time physicists, too, concede that no problem can be brushed aside with the words, "Why, it is quite obvious." Still, in the physicist the desire for unimpeachable logic is not so pronounced and natural as in the mathematician.

A very interesting example.

To support this rather unfavourable thesis, I should like to cite an example which has a direct bearing on the concept of motion.

You may have heard that astronomers have determined conclusively that our Galaxy revolves about an axis through its centre. Popular fiction and even very learned books often mention the rotation of the Milky Way without saying a word about the *frame of reference in which it revolves*. But without a reference frame it is *absolutely meaningless* to speak of the rotation of the Milky Way.

You may well ask how we are to introduce a frame of reference to describe the motion of the Galaxy. To gain an idea of the scope of the problem, imagine the Universe as being comprised of many widely dispersed swarms of bees hovering in "empty" space. Each swarm represents a galaxy. Now try and introduce a likely reference frame. Obviously, it must be associated with some real body or bodies. The only such bodies at hand are our swarms of bees. It is impossible to "attach" a set of coordinate axes to empty space. The only available means of specifying a frame of reference are the swarms of bees.*

We shall not ask how it was discovered that the bees of our swarm — the stars of our Galaxy, that is — take part in a consorted rotational motion within some frame of reference. This would lead us away from our main subject. The

---

*Incidentally, if a bee in a swarm is meant to represent an average-sized star, the picture of the universe preserving the scale will be a rather unexpected one. Thus, if the bee representing the Sun is placed in Moscow, the next closest bee, representing Proxima Centauri, would be located somewhere near Leningrad. The farthest bees of the swarm representing our Milky Way would be removed twice as far as the Moon, while the different swarms of our model would be scores of millions of miles apart. The Earth in our model would be represented by a paltry speck of dust 1/100th of a centimetre across, while a human being to scale would be 10^{-9} cm high — less than the diameter of a hydrogen atom. The scale of such a model is $1 : 10^{11}$.

fact is that no physical experiment conducted on Earth can detect the rotation of the Galaxy, and the conclusion was reached solely on the basis of observations of the relative motions of the stars.

Our concern lies elsewhere.

How was the frame of reference introduced? With what stars — or "bees" — is it connected? How can one hope to "attach" to space the three mutually perpendicular axes of a coordinate system with nothing but a swarm of bees at one's disposal?

The answer to all these questions will be, with your kind permission, highly evasive.

Note that the idea suggesting itself that such a frame of reference must in some way be associated with other galaxies is wrong. Our "mysterious" frame is based on the stars of the Milky Way alone.

We shall not discuss how such a reference frame was introduced. Suffice it to say that this is possible. It is possible to hammer certain imaginary nails into space and attach a coordinate system to them.

Our concern is not so much with how such a reference frame was introduced as with the fact that it had to be defined before any motion (in our case the rotation) of the stars of the Milky Way could be discussed. It is important to realise that the choice of a reference frame is the cardinal problem. Only when there is a frame of reference do the words "the Milky Way revolves" carry any meaning.

With these general remarks of a didactic nature let us return to Newton's laws.

Problem No. 1 in investigating the laws of motion is: "In what frame of reference are these laws formulated?" This first question is probably the most unpleasant one.

Newton found an easy way out. He simply introduced an idealised absolute frame of reference within an absolute space and, accordingly, absolute motion. But, as you will recall, Newton's definition lacks physical content.

A definition, however, is no more than a definition. After all, Newton himself suggested a method of finding "absolute motion" (centrifugal forces) and, consequently, of establishing the absolute frame of reference. If it were as simple as all that the whole problem would be reduced to the correction of an unhappy definition. There would be no cause for worry, Newton's definition would be altered and the absolute frame of reference would remain in mechanics.

The difficulty, however, is that Newton was wrong in principle. *No experiment in the sphere of mechanics can establish the existence of an exclusive frame of reference. This, moreover, is confirmed by the laws of mechanics, the laws of Newton.*

Once again Galileo's relativity principle is proclaimed and investigated.

We shall see this shortly for ourselves. In the meantime let us give up attempts to provide a logically unimpeachable definition of a frame of reference (or of a type of reference frame) for which Newton's laws are valid.

A first attempt to define an inertial frame. It would be useful to remember this definition.

Let us assume that in investigating the motion of bodies we have found a frame of reference in which Newton's laws are true within the accuracy of our measurements. We shall call such a frame of reference an *inertial frame.*

Newton formulated his laws for an idealised absolute frame of reference. We have no idea what this frame is and, furthermore, we won't even discuss whether it exists or not. On the face of it, by introducing the concept of inertial frame all we have done is to replace the word "absolute" by the word "inertial".

Actually, though, our standpoint differs radically from Newton's. We proceed from experience, not from abstract notions. We have established our frame experimentally and named it as we found fit.

Now consider. If there exists in the world a one-and-only inertial frame, then we may legitimately regard it as an absolute frame of reference. But if there exist an infinite number of such inertial frames, then we must concede that *it is meaningless to speak of an absolute system,* at least as far as mechanical phenomena are concerned.

Let us now recall Newton's laws and formulate them as applied to inertial frames.

A preliminary analysis of the first law of mechanics.

The first law: *In an inertial frame of reference every body continues in its state of rest, or of uniform rectilinear motion, unless it is compelled to change that state by forces impressed upon it.*

It is worth noting that the first law of mechanics solemnly proclaims the complete equality of the states of rest and uniform rectilinear motion of a free body in an inertial frame.

It seems obvious that if we introduce another frame of reference, moving uniformly and rectilinearly in respect to the original reference frame, the velocity of the free body will be uniform in the new frame as well. Thus, Newton's first law is exactly the same in the "new" frame as in the "old" one. Conversely, if the motion of the body is described in a reference frame which moves with an acceleration relative to our inertial frame, the behaviour of the body in the accelerated frame will no longer be described by Newton's first law. In such a "bad" frame our free body will not be in a state of rest or uniform rectilinear motion. It will be moving with acceleration.

The conclusion can be drawn: if there has been found a frame of reference in which Newton's first law is valid for a free body, then the law is valid for an infinite number of reference frames moving uniformly and rectilinearly with respect to the primary frame.

Conclusions.

And conversely, there exist an infinite number of reference frames in which the first law is not valid, namely, any frame moving with an acceleration relative to an inertial frame.

The foregoing reasoning may have left you with a feeling of dissatisfaction. After all, we have declared that it is necessary to achieve complete clarity and finality even when we speak of the most obvious things. Therefore, however obvious the assertion, "If Newton's first law is valid in one frame of reference it is equally valid in all reference frames moving

A more precise, but somewhat abstract reasoning in support of our point of view.

uniformly and rectilinearly with respect to our frame", it must be substantiated.

The reasoning should be approximately along the following lines. Let there be given a frame of reference denoted, say, by the letter O. In this frame we are able to describe the motion of bodies and objects by the Newtonian laws. Thus, if the investigated body is isolated and free in our frame it will be either at rest or in motion with a constant velocity V.

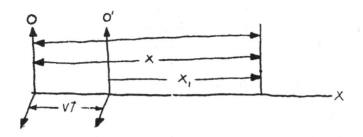

Now we have another reference frame O' which is moving uniformly in a straight line with a known velocity v in respect to frame O.

Our task is to learn how to locate the investigated body in the new reference frame. Before we can speak of the character of its motion in the frame O' we must know its coordinates in that frame at any instant. In other words, we must establish a law which would enable us to go over from one frame of reference to another.

This is fairly simple for the most general case, but we shall examine the very simplest, namely, when frame O' is moving with a constant velocity along the x axis of frame O, and secondly, when the velocity V of our free body is also directed parallel to the x axis of frame O.

Now, if at time $t_0 = 0$ the reference frames were coincident, then at time t the origin of frame O' will have moved away to a distance $S = vt$. Referring to the drawing, we see that the coordinates of a body in the new frame can be found if we know its coordinates in the old frame from the apparent relations:

$$x' = x - vt,$$

$$y' = y,$$

$$z' = z.$$

You may take my word for it if I say that our conclusions are valid for the most general case when the velocities V and v are not parallel to the axes and are not in the same direction.

To return to our example. The coordinates of our body at any given instant in the old reference frame are given by the equations

$$x = x_0 + Vt,$$

$$y = y_0,$$

$$z = z_0,$$

where x_0, y_0 and z_0 are the coordinates of the body at the initial time $t = 0$.

Substituting the equations for going over from one frame to another, we obtain

$$x' = x_0 + (V - v)t,$$

$$y' = y_0,$$

$$z' = z_0.$$

Thus, in the new frame the body is also moving uniformly in a straight line parallel the axis x', but its velocity is now $V' = V - v$.

In other words, we have proved that if Newton's first law is valid in frame O, it is also valid in frame O'.

The reader is invited to look at these formulas again after the Lorentz transformation has been discussed.

Similarly (though from the formal point of view this would be more involved) we can demonstrate that if frame O' is moving *non-uniformly* or *non-rectilinearly* in respect to O, then a body at rest or in uniform motion relatively to O will be in non-uniform or non-rectilinear motion in frame O'.

Very important considerations.

There still remains an important omission in our discourse. In passing from one reference frame to another we tacitly assumed that *the passage of time was the same in two frames*. This follows from the fact that the symbol t in the expression $x' = x_0 + (V - v)t$ must, from its meaning, denote the time measured in frame O. But strictly speaking, if we wish to describe the motion of a body in frame O' we should introduce time t' as measured in the primed frame. Perhaps, at the time of the coordinate readings the body might have been travelling for 5 minutes in the primed frame but only 4 minutes in the unprimed frame! Yet we tacitly assumed that $t' = t$.

Why did we make this assumption? Only because our daily experience tells us that it is correct.[†]

You may well ask what is meant by "time measured in one frame and time measured in another frame". What physical process corresponds to the symbols t' and t and, incidentally, to x' and x? After all, symbols are but symbols. They come to life only when we define uniquely the methods of determining the physical quantities they stand for.

[†]It should be pointed out once again that actually $t \neq t'$. The difference, however, is noticeable only if the relative velocity of the reference frame is comparable with the speed of light.

Thus, the question of transition from one frame of reference to another brings us back to the problem of measuring time. It would be natural at this juncture to provide a recipe for measuring coordinates and time in a given frame.

1. *A coordinate, or length, in a reference frame O is determined by comparison with a measuring rod at rest in the frame.*

2. *Time in a reference frame O is determined by the readings of a clock at rest in the frame.*

In another frame O' we must have a clock and a measuring rod at rest in that frame and perform all measurements *only with that measuring rod and that clock.*

You see that x' and x, or t' and t, actually correspond to different physical processes or measurements taking place in different physical conditions. The notion that $t' = t$ can be accepted only on the assumption that there exist signals capable of travelling with infinite speed.

We shall not go any farther into the jungles of analysis. We have recorded our proposition and explained the meaning of the symbols x and x' and t and t'. That will suffice for the time being.

Thus, the equations for passing from a frame O to a frame O' moving uniformly and rectilinearly along the x axis of the first system have the form:

$$x' = x - vt,$$

$$y' = y,$$

$$z' = z,$$

$$t' = t.$$

This rule, which allows us to find the co-ordinates of events in one frame if they are

known in another, is called the Galileo transformation.

It would be natural now to broaden the
problem. What about the other laws of mechanics? Will they be valid in frame O' if
they are valid in frame O? In other words, is
frame O' also an inertial frame? The answer is
definitely "Yes".

*If O is an inertial frame, then any frame of
reference O', moving uniformly in a straight
line relative to O is also an inertial frame.*

In other words the same idea is expressed
in the statement: the laws of mechanics are
invariant with respect to the Galileo transformation. But if frame O' is moving with acceleration relative to frame O, the laws of
mechanics take a different form.

Gallleo's principle of relativity again.

We thus assert: *There exist an infinite
number of inertial frames which are equally
suitable for describing mechanical phenomena;
the laws of mechanics are the same in all inertial
frames. This is Galileo's principle of relativity,
the fundamental principle of Newtonian
mechanics.*

But let us not be deluded. We have not
provided a water-tight proof of the principle
of relativity. We have only substantiated
the invariance of Newton's first law with
respect to different inertial frames. We have
only proclaimed the invariance of the other
Newtonian laws (which, in fact, we have not
yet formulated). However, if we accept the
Galileo transformation and clearly formulate
the second and third laws, the proof of their
invariance with respect to all inertial frames
becomes a purely mathematical problem. We
shall not go into this. Instead we shall attempt

to get down to the physical meaning of the remaining laws of Newton after which (again and again) we shall return to the first law and Galileo's relativity principle.

The concept of force appears already in the first law of mechanics. The other laws serve in effect to elucidate the concept.

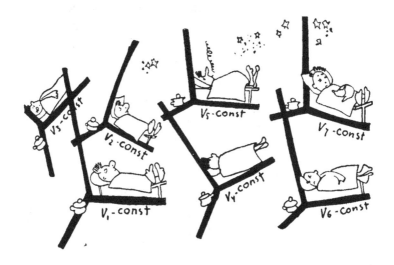

We shall again avoid immaculate definitions and formulas, as any attempt to provide a strict axiomatic definition of the concept of force would lead us into deep waters. We shall only take stock of the salient features.

Generally speaking, force characterises the interaction of bodies.‡ But to say that force

A few words about force.

‡This is not a strict statement of fact, since force may also characterise an interaction between a body and a field. We shall stick to our definition, however, so as not to waste time on a discussion of the involved concept of field, which, besides, is probably the most fundamental in modern physics.

characterises interaction means saying very little. What we want to know is, "How does this interaction display itself?"

To begin with, we can say for sure: if a force is acting on a free body, the body will obtain acceleration. Furthermore, if the same force is applied to different bodies, the accelerations of those bodies will generally be different.

Inasmuch as acceleration, which is a manifestation of force (or interaction), is characterised by magnitude as well as by direction, then force must obviously also be characterised by magnitude and direction. Hence, force is a vector quantity.[§]

You will have probably observed that for the foregoing reasoning to make sense we must know how to measure force, apply equal forces to different bodies, etc.

Measurement of force is based on the assumption that the force acting on a body is proportional to the acceleration imparted to the body:

$$\vec{F} = m\vec{a}.$$

[§]We have mentioned vectors before, but unfortunately we have no time to go into a study of vector analysis. We shall only note the rule of vector composition, known as the triangle or parallelogram rule: To add two vectors, first draw one of them. From its tip draw the second vector. The sum of the two vectors is the vector drawn from the initial point of the first vector to the terminal point of the second.

The statement that "force is a vector quantity" means, therefore, that if there are two forces A and B acting on a body, the net result of their action will be such as if there were a single force C acting on the body. All this is not very precise, but it will do for our purposes.

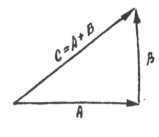

The quantity m — mass — characterises the tendency of a body in an inertial frame to remain, in the absence of an impressed force, in its initial state of rest or uniform rectilinear motion. It is a measure of the inertia of a body.

The question of the quantitative interaction of bodies, in particular, the question "How to apply equal forces to different bodies?", is answered by Newton's third law:

"To every action there is always opposed an equal reaction, or in other words, the actions of two bodies on each other are equal in magnitude and opposite in direction."

$$\vec{F}_{1,2} = -\vec{F}_{2,1}.$$

As for the measure of inertia, mass, it is a most wonderful quantity. Firstly, mass is additive. This means that the aggregate mass of two bodies joined together is equal to the sum of their masses:

$$M = m_1 + m_2.$$

Observations concerning mass in classical mechanics.

You may say that the additivity of mass is as obvious as the fact that the sun rises in the East. But if we start to think carefully we must concede that we have nothing in hand to favour an *a priori* answer. It should be stressed again and again that though the customary usually seems obvious, there is a certain difference between the notions of "customary" and "obvious".

Another, probably equally wonderful, property of mass is that it does not change in passing from one inertial frame to another. This can be expressed in the words: "The mass of a body does not depend on the velocity of the body."[¶]

As a measure of inertia, mass in Newtonian mechanics is completely independent of the physical conditions of a body. We may change its temperature, the pressure acting on it or its location, or place it in an electromagnetic or gravitational field without in the least affecting its mass, or inertia.

Bodies which have nothing at all in common always possess one common characteristic: inertia (mass). On the other hand, Newton's second law provides a uniform method of describing the interactions of the most diverse bodies.

When we consider the motion of bodies with variable mass, Newton's second law takes a more general form:

$$\vec{F} = \lim_{\Delta t \to 0} \frac{\Delta \overrightarrow{(mv)}}{\Delta t}.$$

The quantity $mv = p$ is called the momentum of the body (Newton's "quantity of motion").

If these fragmentary observations have failed to satisfy you, you can cut the knot by assuming mass to be a primary concept and regarding Newton's second law as a definition of force.

If this doesn't suit you either, look up more comprehensive works which deal with the definitions and axioms of mechanics in greater

---

[¶]Running ahead again, we shall note that this statement is only, approximately true. But as long as the velocities considered are much less than the speed of light, the dependence of mass on velocity is quite negligible.

detail.** We shall go no further into this aspect of Newton's laws.

There is one fact, however, not so incomprehensible as surprising, which should be mentioned.

When we discussed the laws of mechanics we took it for granted that everything we said applied to inertial frame. It is now time to ask once again, "What is an inertial frame of reference?"

An attempt to strictly define an inertial frame. A vicious circle.

Earlier we stated our intention to avoid immaculate definitions and be content to verify experimentally whether Newton's laws were valid in a given frame of reference.

But in experimentally verifying, say, the first law, we are confronted with the following problem: how are we to establish that there is no force acting on a body, that it is in fact free. The only logical and precise answer is: we know that there are no forces acting on a body if it is at rest or in uniform rectilinear motion in an inertial frame.

But this answer is unsuited, for that is just what we want to know: *is our frame in fact inertial or not?* Our attempt to define an inertial frame has led us into a vicious circle.

There should be nothing strange in the fact that an attempt to give a logically watertight definition of an inertial frame in terms of Newton's laws, which are themselves formulated for inertial frame, has been unsuccessful.

**It must be said that these things are really not worth our attention, because the whole of Newtonian mechanics can be erected on the basis of several very general principles; the number of postulates and definitions is reduced to the barest minimum, thereby achieving the utmost clarity.

Now our ambitions are much more modest. We no longer care for logic. All we want is to determine purely experimentally and with a sufficient degree of accuracy whether a given frame of reference is inertial or not. In this the best thing to do is to rely upon our intuitive notions of force.

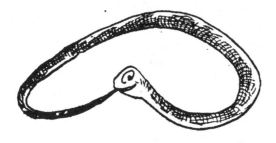

Without laying claim to accuracy, we can say: "If a body is removed 'sufficiently far away' from other bodies and no forces are acting on it, the body is free."

Then, if the body is moving uniformly and rectilinearly or is at rest with respect to a frame of reference, that frame is an inertial frame.[tt]

What is meant by "sufficiently far away"? Well, just very far away. We can state some distance for every specific case.

[tt]The reader who knows about Coriolis forces will readily note that the fact of a free body being at rest is not enough to declare a frame to be an inertial one. Strictly speaking, a single free body is not sufficient to verify the "inertiality" of a frame. It is necessary to investigate the motion of three bodies whose paths, furthermore, do not lie in the same plane. But this is too deep.

These remarks, of course, offer poor consolation. It is hopeless to seek a logically sound definition of an inertial system. We should be satisfied with the consideration that our definition of a free body is at least graphic and physical in nature.

In investigating planetary motion round the Sun, we may hope that the stars surrounding the solar system in no way affect the course of the planets and that the forces acting on the latter are due solely to their interactions with the Sun and among themselves. Having analysed our observations on the basis of this assumption, we find that Newton's laws are valid for a coordinate system referred to the Sun and the sphere of the fixed stars, hence, it is an inertial frame.

True, such a model reference frame is not quite perfect, for the sphere of the fixed stars is not unchangeable. On the contrary, the stars are known to be moving with respect to one another with tremendous velocities of tens and hundreds of miles per second. The relative positions of the stars are constantly changing, but they are so terribly distant that their apparent configurations do not change for many, many years.

The "sphere of the fixed stars" as a standard inertial frame of reference.

If you have ever lain on your back in a field and gazed up at clouds drifting away in the blue yonder you will have noticed that they often seem to be quite motionless. Only when they have disappeared do you realise that they were actually moving; it requires an effort of the imagination to realise that a cloud may actually be travelling at a very high speed.

It is not difficult to calculate the apparent angular displacement with time of a star moving, say, with a speed of 100 kilometres per second on a sphere of 10 light years radius with the Earth in the centre. The values in this example are of the same order as the actual distances to the nearest stars and their velocities relative to the Earth.

Here are some actual angular displacements of stars in the course of a year:

Star	Annual displacement
Barnard	10".27
Kapteyn	8".75
Groombridge-1830	7".04

These three stars have the largest known angular displacements and are unique in this respect. Most stars change their positions by a fraction of a second per year.

It follows then that for unsophisticated reasoning we can assume the stars to be "nailed" to the firmament. In the centre of the stellar sphere is the Sun, with the Earth a short way off coasting round the Sun and turning on its axis (and taking part in several other complex motions).

Incidentally, it is due to the Earth's diurnal rotation that we see the stellar sphere to be revolving as a whole with the relative positions of the stars unchanged.

A reference frame associated to the stellar sphere is inertial. Any other set of axes moving uniformly in a straight line with respect to the fixed stars is also an inertial frame. These are the laws of mechanics. It was impossible for anyone to declare in advance that they would be invariant with respect to the Galileo transformation and that hence every inertial frame is as good as the other. The reverse could well have been the case. Physicists do not create a system of the world, they only register the fact that it is as it is.

Thus they established that the infinite number of reference frames moving uniformly in a straight line with respect to the fixed stars are equally valid for the purposes of mechanics.

Now we may legitimately inquire what is meant by non-inertial frames and what are the laws of mechanics applicable to them.

Suppose we have some standard inertial frame, the fixed stars, for example. We can claim that Newton's laws are not valid in any frame moving non-uniformly and non-rectilinearly with respect to the fixed stars, or in other words, that such frames are non-inertial.

For example, Newton's revolving vessel of water, by means of which he hoped to discover absolute motion, can be used to distinguish a non-inertial frame of reference (the spinning vessel) from inertial frames. But we already know that the experiment cannot indicate a "most inertial" of other inertial frames. Moreover, no mechanical experiment can do so.

The existence of non-inertial frames, on the other hand, easily lends itself to experimental proof. One such frame is constantly at our very feet: the Earth.

As a result of the diurnal (as, of course, the annual) motion of the Earth with respect to the fixed stars, the laws of Newtonian mechanics are not valid in a reference frame rigidly connected with the Earth. True, fortunately for us, the non-inertiality due to the diurnal, and even more so the annual, movement of the Earth is very small. Otherwise Newtonian mechanics would probably have appeared a century or two

Non-inertial frames of reference.

later, for all mechanical phenomena would have been much more complex and Galileo (who, you will remember, didn't even suspect that the Earth was a non-inertial reference frame) would have encountered such incomprehensible phenomena that.... However, that is not our present concern.

That the Earth is a non-inertial frame of reference has been firmly established by a score of different tests. The first was Foucault's famous experiment. If a frame of reference connected with the Earth were inertial, and the laws of Newton were valid in it, a pendulum should oscillate always in the same plane. Actually, the plane of oscillation of Foucault's pendulum shifts gradually.

The essence of Foucalt's experiment is often misunderstood because we intuitively treat the Earth as an inertial frame.

The experiment can best be visualised for the case of a pendulum at the Pole.

With respect to an inertial frame of reference — the fixed stars — the plane of oscillation of the pendulum is constant and passes, say, through the Pole Star, Vega and the Southern Cross. In the course of its diurnal rotation the

Earth shifts away from under the pendulum and a terrestrial observer sees the plane of oscillation turn accordingly. It rotates through 360° in 24 hours and returns to its initial position. A pen attached to the pendulum would trace a rosette like the one shown here (the petals in the drawing are exaggerated).

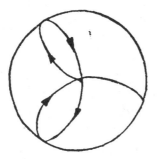 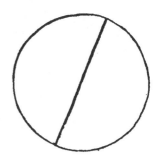

But if we placed under the pendulum a sheet of paper which would be *fixed relative to the fixed star frame*, the pen would trace a straight line.‡ Incidentally, it is quite simple to arrange a sheet in that way. All that has to be done is to make it turn relative to the surface of the Earth in the opposite direction of the latter's rotation with the same angular velocity (one revolution in 24 hours).

If all events on Earth were described in an inertial frame fixed with respect to the stars, then, as all bodies at rest relative to the surface of the Earth would be rotating in such a frame, there would be acting on them a centripetal force

$$F_{cp} = m\omega^2 r,$$

‡In real conditions Foucault's pendulum draws a rosette of different shape, but this is due to aspects of the experiment which do not concern us.

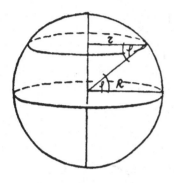

where ω is the angular velocity of rotation and r is the distance to the axis of rotation. Moving from the equator towards the poles, we observe that r becomes smaller, as $r = RCos\varphi$, where R is the radius of the Earth and φ is the geographic latitude of a given point.

Thus, at the equator the centripetal force acting on a body is maximum, at the poles it is nill.

We may well ask where this force comes from. Acting on all bodies lying on the surface of the Earth is the force of gravity. This force is balanced by the reaction of the body's support. The force acting on the support is what we call the weight of the body. A portion of the gravitational force is "spent" on generating the centripetal force which makes bodies move in circles together with the Earth. At the equator this expenditure is greatest and the weight of a body — the force with which it presses on its support — is less than at the poles.

It would be useless to attempt to measure the reduction of weight by means of a scale balance. If such a balance is in equilibrium at the pole it will be in equilibrium at the equator as the weights and the body will lose weight equally. A sufficiently accurate spring balance, however, would immediately show the force acting on the spring at the equator to be less than at the poles.

The angular velocity of the Earth is very small and its effect on weight is slight (the loss of weight at the equator is about four grams per kilogram).

If the Earth revolved, say, twenty times faster, we would experience a wonderful sensation of lightness at the equator. Given a suitable period of rotation, the most unsports-minded person could easily beat all the world records in running, jumping and other events set at the

pole. It would be necessary to specify geographic latitude in conducting sports events. Such an acceleration of rotation, however, would give rise to more serious problems.

To those who care for it a more unusual situation can be suggested. If the Earth revolved fast enough a situation could arise in which, starting from some latitude, the force of gravity would be too small to keep objects on its surface. There would exist a dangerous latitude below which the whole of the gravitational pull would be used up to generate a centripetal force. Any object lying closer to the equator and not fastened to the Earth would immediately fly off into outer space. Crossing the equator would be an outstanding feat for inhabitants of such a planet. (Such a hypothetical planet, however, is hardly of interest, as the first thing to escape from it would be its atmosphere. All the more occasion for us to rejoice that our native planet is so well-conceived.)

As the non-inertiality of terrestrial reference frame is hardly perceptible, Newton's laws can be applied in solving most mechanical problems. There do exist a vast number of problems, however, for which the non-inertial state of the Earth must be accounted for. Thus, in describing the motion of an artificial satellite in a reference frame attached to the Earth, inertia forces must be taken into account, otherwise some astounding absurdities are possible.

In actual life we frequently encounter "very non-inertial" frame. When a bus drives with constant speed down a straight street the non-inertiality of the frame "bus" is due only to the non-inertiality of the terrestrial frame of reference and we do not notice it. But as soon as the driver brakes or accelerates rapidly the bus becomes a "very non-inertial" frame, and the inertia forces jostle us forward or backward.

A rather unexpected remark which the author is very proud of.

Probably neither the bus driver nor the inconvenienced passengers realise that in the final analysis the inconveniences of accelerated motion are due to the fact that the bus brakes with respect to the fixed star frame.

In conclusion, note that if inertia forces
are taken into account, Newton's laws remain
formally valid for non-inertial reference frame,
though their essence changes insofar as it is
necessary to add to the real acting forces certain
inertia forces whose nature is rather obscure.

And now we can raise the question for which
the whole discourse on non-inertial frames was
launched: why is it that the world is so devised

A most curious passage.

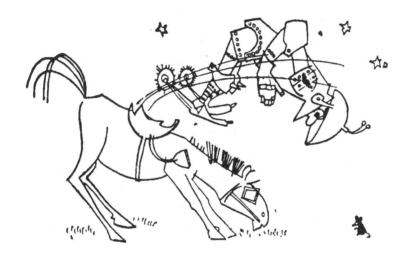

that uniform rectilinear motion with respect to
the fixed star frame has no perceptible effect
on a body while non-uniform motion requires
the application of a force? Or, in other words,
can a reasonable explanation be given for the
existence of non-inertial frames of reference?

At first glance such a question seems to
belong to the same category as queries like

"Why is water wet?" or "Why does a ring have a hole?" This, however, is not the case.

We "attach" Newton's laws to a very definite physical frame: the fixed star frame. You will recall also that we intuitively made the very natural physical assumption that all processes taking place in the solar system in no way depend on the stars. *Only then could we claim that the fixed star frame is an inertial frame.*

The laws of mechanics, we find, are such that any reference frame moving uniformly and rectilinearly with respect to the fixed stars is as good as the other. By no mechanical test can we give preference to any specific frame.

Very well. We may accept this calmly. That is how the world works.

But as soon as we introduce a frame moving with acceleration with respect to the fixed star frame the situation changes radically. The laws of mechanics look different: in such frame we have to introduce some special inertia forces. Furthermore, it is not at all clear why an accelerated frame is any worse (or better, if you like) than an inertial frame of reference. There seem to be no physical reasons why accelerated motion with respect to the distant fixed stars should differ from uniform motion. The very fact that such a difference does exist seems suspicious.

Our intuition tells us that we have come up against something very important, one of the fundamental problems of interest to physicists. All I can say now is that the inequality of inertial and non-inertial frames conceals something strange and remarkable.

An entirely new approach to the problem of non-inertial frame is contained in Einstein's general theory of relativity, but unfortunately we are unable to discuss it here.§§ One more remark on Galileo's principle of relativity is called for to provide a suitable conclusion to our discussion of Newton's laws.

It has been reiterated on many occasions that the laws of mechanics are the same in all inertial frames. Every such frame is as good as the other.

A rather important elucidation of the physical meaning of the relativity principle.

Yet a very simple example seems to refute this assertion completely. Suppose an observer standing on the ground sees a vertically falling stone. An observer looking out of a train moving uniformly on a straight track will say that, in his frame of reference, the stone is falling along a parabola (which is easily proved). The same mechanical action manifests itself in different ways in different inertial frames. Where, then, is the relativity principle?

---

§§It might be pointed out, though, that whereas in investigating the concept of inertial frame within the framework of classical mechanics it was more or less possible to make ends meet by introducing the notion of a body sufficiently far away from other bodies, from the point of view of the theory of relativity this approach is unsuitable. What is more, the words "sufficiently far away" simply mean nothing.

Two bodies may be "very far" apart when the distance is measured in one frame of reference, and "very close" in a reference frame moving uniformly and rectilinearly with respect to the first one (this will be more readily understood by those who hold out till Chapter XIII).

So we find that the state of affairs with the concept of inertial frame is even worse than had seemed at first. True, Einstein's works have served to clarify the problem somewhat, but we shall not discuss them.

But there is no contradiction here. The relativity principle does not demand that some particular physical process (in our case the falling of a stone) should look the same in different inertial frames.

Take an experiment staged in some inertial frame. For instance, we are investigating how a stone falls to the ground.

Now we repeat the experiment in another inertial frame with *all* the conditions exactly duplicated in the new frame.

By "duplicating all the conditions" we mean, among other things, that the initial conditions in the new reference frame must be the same as in the old one.

In the first case, when the reference frame was fixed to the Earth, the stone had *no* horizontal velocity at the initial moment. But in the frame attached to the carriage there *was* a horizontal component of velocity at the initial moment. Therefore, the description of the experiment was not the same in two frames. If the experiment with a falling stone is repeated in exactly all details in the railway carriage then, the relativity principle asserts, everything will take place as on Earth.

A stone with no horizontal velocity with respect to the carriage at the initial velocity should fall vertically down relative to the walls of the carriage, and the law of fall must be the same as when it falls on the ground. And this, of course, is what we actually observe.

This consideration is frequently overlooked, which leads to misunderstandings: complete identity of the conditions of an experiment in two inertial frames presumes that the initial conditions must be the same there.

In mechanics there is no "exclusive" frame of reference, and every inertial frame is as good as the other.

But maybe experiments in some other physical field, say, with light or electromagnetic waves, will make it possible to establish the existence of such an exclusive frame?

Résumé and more doubts.

This question remained open till 1905, when Einstein enunciated his special theory of relativity. In fact, it was the solution of this problem that led him to the development of his theory.

CHAPTER VI,

which, the author hopes, is
rather interesting

NEWTON. GRAVITY

Introduction. Discourse
and apologies.

The chapter on gravity deserves to be opened
with a flourish. "The most mysterious and
unexplored phenomenon of nature." Which
is true, for Newton merely established its
existence while Einstein only lifted a corner of
the curtain of mystery shrouding it (the general
theory of relativity).

Gravity is intrinsic to all bodies without
exception. The properties of gravity are uniform
throughout the whole of the observable part of
the universe. Gravity is omnipotent. It is impossible to escape from it or enhance it. In
short, gravity is gravity.

Gravity is probably the only physical phenomenon which we are incapable of influencing
to the slightest degree.

True, there have been some press reports to the effect that the possibility of controlling gravity was being studied and that some results had actually been obtained. It is hard to say, however, whether such reports are the fruit of an overzealous reporter's imagination or distorted information about actual researches.

The appearance of this chapter in the book is another minor "mystery" in the story of gravity. Unfortunately, we shall not discuss the general theory of relativity and therefore everything relating to gravity will remain outside the scope of our discourse. The only justification of this chapter is that its subject matter is indirectly connected with the special theory of relativity.

This refers especially to the part which discusses whether gravity propagates with infinite or finite speed. An analysis of this question (like the analysis of the fundamental concepts of mechanics) is a desirable preliminary for a sober acception of the ideas of relativity theory.

For, if I may say so again, one of the main obstacles to an understanding of Einstein's theory is the profound subconscious conviction that the basic concepts of classical mechanics are absolute and, as it were, preordained.

The author hopes to convince the reader that all that follows will be useful for a better understanding of the theory of relativity.

And, conversely, once it is realised that physics is based on experiment and does not recognise *a priori* concepts, and once a clear idea is gained of the hypotheses on which classical physics is based, Einstein's theory should seem no less natural and no more difficult than classical physics.

The main justification, however, is that the wonder-phenomenon of gravity is associated with some of the most remarkable physical theories. It would not be an overstatement to say that probably no other findings in the whole realm of physics can match the theories enunciated by Newton and Einstein.

Finally, there is probably no better example in the history of science of a scholar's work on a seemingly hopeless problem, persistent work which after many fruitless years was ultimately crowned with such brilliant success.

In short, it would be difficult to provide a better example of the triumph of "justice supreme" in science.

By 1666, Newton already had his theory clearly in his mind's eye. Strangely enough, but it seems that the famous apple anecdote is quite true. Newton himself remarked much later that the idea of the existence of a uniform force which made all bodies without exception gravitate towards one another occurred to him when he observed an apple falling to the ground. Be that as it may, but the idea itself is remarkable.

Newton had at his disposal a motley array of facts many of them seemingly contradictory.

For one, he knew the laws of planetary motion discovered empirically by Kepler. It took Kepler twenty-five years to piece together countless data on the apparent positions of planets and detect the hidden laws. He carried out a formidable amount of work to discover the laws, but was unable to explain them, although he did venture the idea of a gravitational force.

That all bodies on Earth tend to fall down needs hardly to be stated. The difficulty is that the force with which the Earth attracts bodies seems to be constant and, as far as Newton could judge, independent of a body's distance from the centre of the Earth (for in Newton's time experimental techniques could not detect a change in the weight of a body lifted one or two miles above sea level). The laws of planetary motion, on the other hand, are such that the hypothetical force which Newton thought to govern them should definitely change with distance.

Furthermore, if there exists a uniform gravitational force, why don't bodies on Earth gravitate towards each other? The experimental data, you see, is rather confusing.

True, the idea of a uniform gravitational force was hovering in the air, and Newton had his predecessors. But none of the exponents of gravity was able to give a quantitative explanation to the laws of planetary motion or answer the objections of their opponents.

Robert Hooke, one of the most vivid and original scientists in the annals of physics, had all but discovered the law of gravitation. In his *"Attempt to Prove the Motion of the Earth from Observations"*, written in 1674, he says that his "system of the world" depends upon three suppositions. "First, that all celestial bodies whatsoever, have an attraction of gravitating power towards their own centres, whereby they attract not only their own parts, and keep them from flying from them,... but they do also attract all the other celestial bodies that are within the sphere of their activity.... The second supposition is this, that all bodies whatsoever

An outline of the basic propositions of the gravitational theory. The author wishes to show the formidable difficulties overcome by Newton. In passing he goes into a bit of "psychoanalysis".

The author is happy to say that in the present case all claims to literary style must be addressed to Robert Hooke.

that are put into a direct and simple motion, will continue to move forward in a straight line, till they are by some other effectual powers deflected and bent into a motion describing a circle, ellipsis, or some other compound curve line. The third supposition is, that these attractive powers are so much the more powerful in operating, by how much the nearer the body wrought upon is to their own centres."

He goes on to say that he had been unable to ascertain by experiment the various degrees of attraction, but if his theory were developed further, astronomers would be able to establish the law according to which all celestial bodies move.

Hooke then goes on to observe that he himself is much too busy with other problems and that the best course would be for someone to develop his ideas further.

The closest investigation of archives would hardly give food for authoritative assertions, but still it seems that Hooke was not altogether sincere. He must have been only too well aware of the overwhelming importance of the problem which he had studied for many years. It is more likely that he gave up his search for an answer not so much because of pressing engagements as simply because he was unable to solve the problem.

It is a long way from Hooke's qualitative speculations to Newton's laws. We can understand the latter's vociferous indignation when Hooke suggested (true, with proper discretion) that he, too, had had a hand in the discovery of the law of universal gravitation. (I refer those who are interested in the priority controversy between Newton and Hooke, as well as in other such squabbles in which the great Newton was involved, to sundry biographical dissertations to which I have nothing to add.)

It must be said that the public usually displays a heightened interest in such cases and that, regretfully, scientists of all times and nations all too often provide more than enough food for controversy. In the Hooke–Newton dispute, however, there is an interesting psychological point.

Newton's biographers all agree that in his latter years Sir Isaac developed a most quarrelsome disposition. Imperious, proud and quick to take offence, he furthermore begrudged other people their desserts. This is as it may be, but his behaviour can hardly be attributed to proudness. In his work, even overlooking for a moment his genius, Newton was always first and foremost a scientist in the best sense of the word.

This is seen in his extreme exactingness towards himself. And, naturally enough, he extended this exactingness to the work of others.

If we recall that, in Newton's own words, he had formulated the basic ideas of the gravitational theory in his own mind by 1665 but had refrained from publishing them because he regarded them as raw material which a true scientist should keep to himself, his reaction to Hooke's claims is understandable.

On the other hand, though, we can understand Hooke's chagrin when his ideas were rated so poorly. After all, he *had* voiced the idea of a gravity force, and moreover, he had predicted that it should be in inverse proportion to the square of the distance. This would seem sufficient to merit him glory and fame and recognition, but not in Newton's view.

Newton measured things by his own yardstick and he can hardly be blamed for considering, quite sincerely, that Hooke's ideas were self-evident and, furthermore, too vague to merit publication. Of course, he had no right to gauge other scientists by his own standards, but that is another matter.

With all his shortcomings, it should be remembered that for decades Newton was reluctant to publish such discoveries of his as the calculus or his considerations on universal gravitation. This hardly suggests that he was very much concerned with immortality.

Academician Vavilov, however, questions the truth of this.

There is a story according to which Newton had arrived at the analytical formula of the gravitational law as early as 1666. But an attempt to describe the Moon's motion by means of the gravitational law was unsuccessful, because Newton was in possession of erroneous experimental data concerning the dimensions of the Earth. As a result, the value of the acceleration of gravity at the Earth's surface as calculated on the basis of lunar motion, disagreed with experimental data. Only in 1682 did he learn that new data on the length of the meridian had been obtained.

Newton, they say, was so excited that he was unable to carry out some very simple calculations, which he asked an unknown friend to do for him. And that was how the gravitational law was finally formulated.

In the final analysis, it does not matter whether Newton was right or wrong in his squabble with Hooke. The important thing is that no one but Newton possessed a sufficient command of mathematics and physics to enable him to deduce the empirically established laws of planetary motion from a general law of gravitation. Neither could anyone solve the reciprocal problem of formulating the law of interaction of bodies from Kepler's empirical laws. This was done in the *Principia*.

In Newton's hands the gravitational law provided answers to all the major questions connected with the motion of celestial bodies.

But this is not all, the force of gravity calculated by means of the gravitational law coincided very nicely with experimental results. One could hardly have demanded more convincing proof, and yet almost a century passed before the gravitational theory won final recognition in the scientific world.

To Newton's contemporaries, the gravitational theory was more revolutionary and astonishing than the theory of relativity seems to us. One of the reasons for this is that scientific standards in the seventeenth and eighteenth centuries were much lower than in our time. Not that there were fewer talented scientists. Simply the medieval outlook still weighed heavily on the most brilliant minds of the age. Newton himself was a dedicated interpreter of the Gospel. What could be expected of others?

An interesting historical fact.

Remember that his contemporaries still clung to the traditions of the physics of hypotheses and you will understand their reaction when an analytical law of interaction of bodies was formulated to describe the most fundamental and intrinsic property of all bodies. To the scientists of those days this seemed almost an affront.

It is hardly surprising, therefore, that even such men as Leibnitz, Huygens, Euler and Lomonosov rejected the idea of gravitation.

Leibnitz, for instance, regarded it as "a chimerical thing, a scholastic occult quality." Huygens wrote that he could not accept Newton's explanation of the cause of tides, "nor other theories of his which are founded on the principle of gravitation, which I thought absurd".

Opposition to Newton was especially great in France, where scientists were under the spell of the teachings of René Descartes.

It is not for us to appraise the scientific impact of the great French philosopher's ideas. We need not have mentioned here his explanation of the apparent motion of celestial bodies if not for one aspect of his theory of matter which is of interest to us. For it is in that theory that the mysterious substance called the ether first appears. The ether which engaged the attention of physicists up to the twentieth century!

First, but by no means last, mention of the ether.

According to Descartes, the ether was in constant turbulent motion involving all the planets. In the course of this motion, the components of matter, which were initially in a chaotic state, were divided into three types of particles.*

The first, and coarsest, are particles which make up the Earth, planets and comets.

The second, finer, particles went into the making of the Sun and stars.

---

*To use an extremely profound analogy, we might say that Descartes postulated something of a universal separator.

Finally, the third type are infinitely fine particles.

According to Descartes, the interaction of heavenly bodies is due to the pressure of the ether on them. This pressure is transmitted through the ether from one body to another, and thus celestial bodies act on each other.

Note that in order to transmit an action (or force) over a distance there had to be a medium possessing certain mechanical properties, the ether.

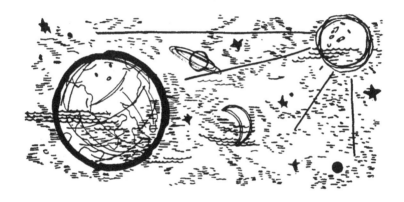

Descartes and his disciples sought to explain gravitation by means of a model which reduced everything to the action of bodies on the ether and the reciprocal action of the ether on heavenly bodies.

It goes without saying that Descartes was unable to develop an analytical expression for his theory. Scientists were attracted by the charm of clarity and obviousness in his hypothesis.

A scathing characteristic of the scientific atmosphere of the age was given by Voltaire, who made a hobby of physics in his youth:

"A Frenchman who arrives in London will find Philosophy, like everything else, very much changed there. He had left the world a *plenum*, and now he finds it a *vacuum*.'

"In Paris the universe is filled with etheric turbulence, here invisible forces are acting in the same space.

"In Paris it is the pressure of the Moon on the sea that causes tides, in Britain, however, it is the sea that gravitates to the Moon.

Descartes often signed his name Cartesius. Hence, the adjective Cartesian.

"The Cartesians explain everything by means of pressure which, frankly speaking, is not very clear; the Newtonians achieve everything by means of attraction which, however, is not much clearer.

"Finally, in Paris the Earth is considered to be elongated at the poles like an egg, in London it is compressed like a pumpkin."

This was written in 1727, forty years after the *Principia* had been published, yet Voltaire's scepticism extends equally to the theories of Newton and Descartes.

This illustrates the obstacles that the gravitational law had to overcome before it drove home. Still, slowly but surely the truth triumphed, and by the beginning of the nineteenth century all doubts as to the truth of Newton's laws were dispelled. Characteristically, it was the French scientists of the latter half of the eighteenth century who gave celestial mechanics its final polish and demonstrated that the gravitational theory was true and that there was no truth without it.

The gravitational law is probably the pinnacle of the method of principles. It says nothing of the why and wherefore of gravitational action. All it says is how this mysterious force acts:

And finally, the law of gravitation itself.

$$F = f \frac{m_1 m_2}{r^2},$$

where F is the force of mutual attraction between two bodies, m_1 and m_2 are their respective masses, r is the distance between them, and f is a universal constant, equal to the force with which two bodies of unit mass attract each other through unit distance, called the gravitational constant. In the CGS system

$$f = 6.67 \times 10^{-8} \frac{cm^3}{sec^2 \cdot g}.$$

The infinitesimal value of f explains why we fail to observe gravitational forces between terrestrial objects.

Three curious features in Newton's law should be mentioned.

There is an apparent analogy between gravitational forces and electrical charges, which are interactions of an entirely different nature:

Feature No. 1

$$F \propto \frac{e_1 e_2}{r^2},$$

which is Coulomb's law.

We shall not go into an explanation of this interesting coincidence. (True, there is one fundamental difference: gravitational "charges" have only one sign.)

Newton's law presumes, and we shall dwell on this at greater length, that gravity propagates with infinite speed. Under the gravitational law, in order to determine gravity forces at any *specified* instant it is sufficient to know the distance between the respective bodies at *that instant*. How the distance changes with time, or to put it scientifically, the space-time

Feature No. 2

relationships between the interacting bodies, are immaterial.

Let us see how Newton's law changes if we assume that gravity propagates with finite speed, all other relationships remaining the same.

Suppose we have two bodies interacting according to Newton's law, the gravity forces propagating with a finite velocity c. If the bodies are at rest nothing changes. Not so if they are moving relative to each other.

The first question, of course, is: What do we mean when we say gravity propagates with a finite speed equal to c? In what frame of reference? Therefore we must accept as a premise some "absolute frame" in which the velocity of gravity is c.

We don't know and we don't care why the velocity of gravity is finite; perhaps bodies constantly emit gravity waves travelling with finite speed through space. Our concern is to see how Newton's law changes in such circumstances.

To simplify matters, let us consider the case when one body is at rest within our "absolute reference frame". At the initial time $t_0 = 0$, let the second body be moving uniformly towards the first with velocity V. When the bodies are at rest the interacting force is defined by Newton's law:

$$F = f \frac{m_1 m_2}{r_0^2},$$

where r_0 is the distance between the bodies at rest. At some time t_1 the distance between the bodies becomes $r_t = r_0 - Vt$.

What is the interacting force in this case? As the velocity of gravity is finite, the interaction between the bodies will be determined *by the distance between them not at the given instant but at some earlier moment.* The "wave" of gravity which reached the first body at time t was emitted by the second body at some earlier instant $t_1 < t$. This time is easily determined, but it is probably not worthwhile going into equations. All the more so as we have failed to mention something more important.

For actually we have not indicated the reference frame in which the velocity of gravity is considered, and as long as we have no reference frame it is meaningless to speak of the velocity of gravity.

Obviously, if such an absolute frame exists it must be connected not with any two random bodies, as in our case but with the properties of space itself (the fixed star frame?).

Well, perhaps now we can find our absolute frame? *Incidentally, how are we to determine the velocity of gravity in other reference frames?*

Attention! This question is not so naive as it seems.

In short, as soon as we assume that gravity forces propagate with finite speed the general picture gets much too involved and the equations of motion of celestial bodies become very complicated.

Newton made short work of all such difficulties by assuming the velocity of gravity to be infinite. Thereby he introduced action at a distance. Frankly speaking, it is rather difficult to accept this idea. Our very being revolts against it. All known actions propagate with finite speed, even light. Gravity seems to be a strange exception. Well, all the more cause to wonder at Newton's genius and intuition.

Running ahead, we can say that today, after Einstein, we know that Newton was wrong. The speed with which a gravity field propagates is finite and is equal to 300,000 kilometres a second. Furthermore, this velocity possesses one strange quality: it is constant for any reference frame and does not change from one frame to another.

Thanks to the tremendous speed of gravity, the corrections to Newton's law due to gravity "arriving late" are minute. It is hardly surprising, therefore, that for two centuries the absolute validity of the gravitational law was unquestioned. And this brings us to the next feature.

Feature No. 3

Undoubtedly, most wonderful in Newton's law is that *the gravity force is completely determined by the inertial masses of bodies.*

In no way does gravity depend on chemical composition or electrical charge or physical state. It depends only on mass, or ultimately, on the inertia of the gravitating bodies.

Intuitively we feel that there must be some deep-lying connection between inertia and gravity. Yet gravity and inertia are such apparently different properties that physicists felt it necessary repeatedly to check experimentally whether mass as determined by the laws of mechanics (inertial mass) and mass in the gravitational law are actually the same.

Here, too, Galileo was the first.

The fact that in the Earth's gravity field all bodies fall with the same acceleration is the main proof of the equality of inertial and gravitational mass.

Let us check this. Acting on a body of mass m in the Earth's gravitational field is a force

$$F = f \frac{m_g M_g}{r^2},$$

where m_g is the gravitational mass of the body determined from the gravity law, M_g is the gravitational mass of the Earth and r is the distance to the centre of the Earth.[†]

Without assuming in advance the equality of gravitational and inertial mass, and applying the second law of mechanics, let us find the acceleration of the body in the Earth's gravitational field:

$$g = f \frac{M_g}{r^2} \frac{m_g}{m_i},$$

where m_i is the inertial mass of the body and g is the acceleration of gravity.

In this equation the multiplier $f \dfrac{M_g}{r^2}$ is the same for all bodies on the surface of the Earth. The second multiplier, $\dfrac{m_g}{m_i}$ may, we have assumed, change depending on the nature and composition of the physical bodies concerned.

But since in the Earth's gravitational field all bodies fall with the same acceleration $g = 9.81$ m/sec^2, we conclude that the ratio of the

---

[†]It can be easily demonstrated that a sphere (the Earth) attracts bodies as if the whole of its mass were concentrated in the centre.

gravitational mass to the inertial mass $\dfrac{m_{\mathrm{g}}}{m_{\mathrm{i}}}$ is the same for all bodies regardless of their physical nature.

Hence, the gravitational mass of a body is determined completely by its inertial mass and can be taken simply as equal to the inertial mass, provided the units of measurement are appropriately chosen.

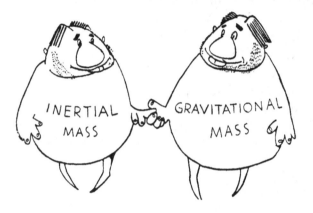

Thus, the gravitational attraction of bodies depends on their inertia, and only on their inertia. This should be regarded as a very unexpected conclusion. For to expect such a connection merely on the basis of general considerations would be as right as, say, to believe that the positions of the planets and stars at the time of one's birth can determine one's fate.

However, unlike the propositions of astrology, black and white magic and other occult sciences, the fact that the gravity of bodies is wholly determined by their inertia rests on the unshakable foundation of exact experiment. Furthermore, the surprising result induced physicists to verify it experimentally again and again.

Newton himself checked Galileo's findings by investigating the motion of pendulums made of different materials.

In 1828, the German mathematician and physicist Friedrich Bessel investigated a wide range of substances in a similar way and proved the proportionality of inertial and gravitational mass to an accuracy of $60,000^{-1}$.

The summit of experimental precision was achieved by the Hungarian Eötvös and his associates in a series of experiments carried out between 1896 and 1910. The proportionality of inertial and gravitational mass was established by them with the incredible accuracy of 5×10^{-9}.

Six years later, in 1916, Einstein finally formulated his general theory of relativity, in which he described a most ingenious method of investigating the mysterious gravitational theory.

At this juncture it would be appropriate to say a few words about Albert Einstein, the scientist and the man. So much has been written about him that we hardly need to go into a detailed account of his life.

Like Newton, in science "his faculties excelled the whole human race", and Newton's is just about the only other name that can be ranked side by side with his. Einstein's views outside the realm of science are an example of genuine militant humanism, humanism in the loftiest sense of the word.

He was a most peaceful man who wanted nothing for himself but the opportunity to work. Intense mental work, an ardent desire to find yet another "smoother pebble or prettier shell" were such intrinsic qualities of his, so inseparable from the very substance of his being, as to be even beyond wonder. That inexplicable accumulation of qualities which is usually defined as genius gave rise in him to an urge to work that was practically instinctive. And perhaps even more to be admired in Einstein is his wonderful integrity and innate probity as a physicist and as a man. His exceptional talent would have made it possible for him to achieve outstanding results in all spheres of physics without overexerting himself in the least and enjoy the fruits of scientific achievement. But he spent the last thirty years of his life working on a problem which he considered to be most important, even at the cost of departing from the smooth highway of contemporary physics. As he wrote himself, hope had often betrayed him and as many times he had to partake of the bitter fruit of frustration.

He worked practically alone. The interests of other scientists lay in other spheres. Perhaps there are not a hundred physicists in the world today capable, without a good deal of preliminary study, of describing the purport of the work of his latter years. (In quantum mechanics there may be several thousand such scientists.) I find it difficult to cite another example of such intellectual purposefulness in the whole history of science. It is not for us to judge of Einstein's achievements in those years. But even if he had not been the author of relativity theory, even if he had not obtained in those years such incidental results in other spheres which alone would have been sufficient to immortalise his name in physics, even then they could have made him famous.

The inscription on Newton's tombstone in Westminster Abbey is simple and reserved:

"Mortals, congratulate yourselves that so great a man has lived for the honour of the human race."

These words could be applied with even greater justification to Albert Einstein.

We are today convinced of the approximate validity of Newton's law and that Einstein's gravitational theory offers a further approximation of the truth, even though it has not yet run the gauntlet of verification by experiment ordained to every physical theory.

Countless experiments confirm Newton's law. The general theory of relativity has so far been confirmed by only four effects.[‡] These are: the equality of inertial and gravitational mass, the continuous precessional motion of the perihelion of the orbit of Mercury, the deflection of light in a gravitational field, and the change in the oscillation period of atoms in a gravitational field (the redshift).

Nevertheless, the very fact that Einstein predicted and explained these amazing effects is in itself proof of the validity of the general theory of relativity.

We do not know how the theory will develop and change, but it is already apparent that the cardinal problems of future physics are connected with the gravitational field theory as we know it today.

[‡]This text was written about 60 years ago. Today we have many more confirmations of the validity of general relativity. The existence of black holes is an exciting confirmation. And probably still more exciting was the discovery of gravitational waves in 2015 — A.S.

CHAPTER VII,

which, though rather vague, after many digressions finally explains why physicists were so attracted by the ether hypothesis

LIGHT, THE ETHER (Newton, Huygens)

...There is no evidence for its existence, therefore it ought to be rejected.

NEWTON

Any attempt to pinpoint the branch of physics (or any other science, for that matter) which could be said to have played the greatest part in its development is inevitably academic and somewhat scholastic.

First, some history.

All that can be said for sure is that light (and later on electromagnetic phenomena in general) have always been a subject of interest to physicists. Electricity and magnetism have, in one way or another, been in the front line of physical research. It was the investigation of electromagnetic phenomena that led to the development of the special theory of relativity and quantum mechanics and, it is worth remembering, such technological achievements which have served to transform completely the very life of human society.

The pioneers along this road were Newton (Newton again!) and Christian Huygens (1629–1695).

A contemporary of Newton, and as a physicist undoubtedly second to none but him, Huygens left a trace in many branches of physics.

To his credit, besides his classical works in optics, are excellent studies in astronomy and, especially, mechanics. To him we are indebted for the first accurate clock (the pendulum clock), an invention that can well be placed in one line with, say, the development of jet aircraft.

Newton himself spoke of him as "the illustrious Huygens", and the President of the Royal Society and first physicist of the world was none too liberal with compliments.

Since we are concerned with light, a brief historical outline of the study of light is appropriate, the more so it contains some interesting and amazing facts.

What information we have concerning the study of light by the Greeks and Romans is fragmentary but nevertheless interesting. The ability of Greek and Roman philosophers to develop extremely complex and subtle speculative theories is well known. On the other hand, equally well known is the deliberate scorn for experiment in ancient science.

Opticians of Antiquity.

Nature chastised the ancients for this, and we can only wonder at the sad state of physics in those days as compared with mathematics.

In optics, it seems, the situation was somewhat different. At least, there were a number of theories, all of them highly hypothetical. The atomistic theory of light of Democritus and Epicurus was no exception. All theoretical deductions were abstract and speculative and unsupported by experiment. Still, there are indications that after all the Greeks may have not been so disdainful of experimental physics.

Ptolemy's treatise on optics, we find, gives angles of refraction of light at air-water intersurfaces to a high degree of accuracy. It seems that Ptolemy must have conducted some experiments.

Roman historians tell us that the myopic Emperor Nero used a polished emerald to improve his sight. Thus we find that the principle of the eye-glass, an instrument of unique application, was known in ancient times.

Finally, there is the famous story of Archimedes using mirrors to set fire to Roman ships besieging Syracuse, which also testifies to experimental work in the sphere of optics. The story itself may well be a pure invention, but it could appear only from a knowledge of the focusing properties of concave mirrors.

Scores of other reports seem to indicate that many experimental data in optics (especially geometrical optics) were known to the Greeks.

All this might suggest that we are not so well informed of the state of science among the ancients. Nevertheless, it is assumed to be an established fact that physics did not exist as an experimental science in antiquity.

Interest in optics was revived in the early years of the Renaissance. Spectacles were invented (or rediscovered?). Leonardo da Vinci expressed some really brilliant ideas in scattered notes. Interesting works of other scientists appeared, but all these were nothing more than random ideas.

The turning point came in the early seventeenth century and again it is associated with Galileo Galilei.

Galileo and optics.

It is not that Galileo enunciated a new and orderly theory of light phenomena. He considered light to be a flux of minute indivisible particles, but that was not new.

Galileo built excellent optical instruments, but men had done so before him.

He has some interesting observations concerning various problems of physical optics (phosphorescence, for example), but they are fragmentary and dispersed and they played no great part in advancing the science of light.

But in optics, as in mechanics, Galileo was the first to apply consistently a new method of investigation. And in optics, as in physics in general, he was first and foremost an experimenter.

It was this new approach to scientific problems that led him to the momentous question: "How fast does light travel?"

Rather, it is not the question itself that is so momentous as the way in which he formulated it. Galileo doesn't speculate endlessly whether and wherefore light should propagate with finite or infinite speed. He leaves such problems to others. His ideas are clear-cut and to the point: "Is it possible to devise an experiment to determine the velocity of light?"

The problem is debated by our old friends Salviatus, Sagredus and Simplicius in Galileo's famous *Dialogue*, his last and greatest work. "Everyday experience," Simplicius says, "shows that the propagation of light is instantaneous.... The flash of artillery fired at great distances is observed at the same time, unlike the sound which reaches the ear after a considerable interval."

Sagredus: "But Signior Simplicius, the only thing I am able to infer from this familiar bit of experience is that sound travels more slowly than light; it in no way convinces me that light propagates instantaneously and does not take some time, however small. Neither am I more impressed by another observation which is expressed in the words, 'When the Sun appears above the horizon its rays immediately strike our eyes.' For who will prove to me that in fact the rays do not appear at the horizon before they reach our eyes?"

Salviatus (who expresses Galileo's opinions) then describes an experiment, which Galileo himself had evidently conducted in an unsuccessful attempt to determine the velocity of light. The principle of Galileo's experiment was along much the same lines as later experiments performed to determine the velocity of light in terrestrial conditions.

Two observers equipped with lanterns take up positions at a distance of several miles from each other. The first opens the shutter of his lantern at time t_0. As soon as the second observer sees the light he opens his shutter, and the first one registers the time t_1 when he sees the light flashed back by his partner.

Galileo attempts to determine the velocity of light.

Assuming that light travels with the same speed in all directions and knowing the distance r between the observers, we have for the velocity of light:

$$c = \frac{2r}{t_1 - t_0}.$$

Today we know that at best this experiment makes it possible to determine the reaction of the observers, not the velocity of light. Galileo did not realise the tremendous speed of light.

Naturally enough, he did not stop to consider how the velocity of light changes in passing from one reference frame to another; a question which later worried physicists for upwards of two centuries. This would be asking too much of him. As it is, Galileo was the first to approach the problem as a true physicist: careful experiment first, then theoretical deductions.

After Galileo we must name his compatriot Francesco Grimaldi (1618–1663). A teacher of rhetorics, and later mathematics, at a Jesuit college in Bologna, he devoted his life to the study of optical phenomena.

Grimaldi's name is not very famous in the history of science. He was incapable of sweeping theoretical generalisations and was frequently unable to offer satisfactory explanations of his own observations. These limitations of his may be due to the fact that he was an exemplary member of the Jesuit Order and all his life opposed the ideas of Copernicus and Galileo. He was an outstanding experimentalist, however. Suffice it to say that he discovered light interference and diffraction and the dispersion of sunlight by a prism. His theoretical ideas already contained some elements of the wave theory of light. Unfortunately for Grimaldi, however, Newton's work on optics were so much superior to his that it is hardly surprising that they were largely forgotten after Newton's *Opticks* appeared.

Particles or waves?

Prior to Newton the theory of light phenomena was investigated by Descartes and, especially, Hooke. Descartes, however, was more of a mathematician and philosopher than a physicist, while Hooke usually never pursued his generally brilliant ideas to the end.

The first theory of light phenomena worthy of the name was advanced by Newton. You will recall that in mechanics, too, he introduced certain hypothetical propositions, though in veiled form.

In optics hypotheses are essential. Optical phenomena are of too varied a nature to allow for a few uniform principles. A hypothesis is necessary to bind together the available facts.

The facts on hand in Newton's time pointed to either of two possible theories:

corpuscular, according to which light is a stream of particles;

wave, according to which light is wave motion.

Newton inclined rather to the former, Huygens expounded the latter. By the beginning of the nineteenth century the controversy was resolved in favour of Huygens and there seemed to be no more doubt concerning the wave nature of light.

Twentieth-century physics has rehabilitated Newton.

It is probably worth recalling the fundamentals of wave motion, as superfluous observations in day-to-day life may lead to considerable misconceptions.

Throwing a stone into a pool of water and observing the waves spreading out in circles, we are usually content to say that they are moving with such and such speed. We can even measure that speed without stopping to consider what actually is being transferred by the motion and how the particles of the medium through which the wave is propagating are behaving.

Wave motion is a process in which energy is transmitted through a medium. The particles of the medium oscillate about positions of equilibrium.

Something of a definition.

The propagation of a wave consists in that more and more particles of the medium begin to oscillate, the direction of their motion not necessarily coinciding with the direction of the wave.

An analogue of longitudinal wave motion is the "grapevine telegraph". A report originating at some source reaches the other end of the line through a number of participants.

The "grapevine tele-graph", though, is rather an analogue of wave motion through a damping and distorting medium.

An even closer analogue is the relay communication system used in ancient times. Several dozen runners are posted along a route. A message is handed to the first runner, who delivers it to the next one and returns to his post; the second runner hastens to hand on the message to the third, etc. Such a relay illustrates a longitudinal wave; the message is the "transmitted energy", the messengers are the "particles of the medium".

In a longitudinal wave the particles oscillate parallel to the direction in which it propagates.

It is not difficult to find a familiar analogy to transverse wave motion, in which the particles of the medium oscillate perpendicular to the direction of propagation.

Like most analogies, this one offers a very approximate illustration of wave motion.

Imagine a flock of birds sitting in a long row on a wire. A bird at one end flutters up at a false alarm and, seeing that all is quiet, resumes its perch. Its neighbour flutters up a fraction of a second later, and so on down the line. By the time the false alarm signal reaches the other end, the first birds will have completely forgotten the matter. In this case the alarm is the transmitted "energy" and the birds are the "particles" moving perpendicular to the direction of the signal.

Thus, the particles of a medium through which a wave is passing move about positions of equilibrium. If an excitation has developed at one point, a wave may spread only if the particles are in some way interconnected. This is obvious.

Less trite is the following observation: a wave will propagate without distorting or damping only if the forces acting between the particles are of a very specific nature, namely, so-called elastic forces. Such a medium is called absolutely elastic and is, of course, an idealised case. Yet there are many substances, lying as wide apart as steel and air, through which waves travel with very small loss of energy.

That air is a very elastic medium we know from the fact that we can hear a person speaking scores of yards away. The small energy of our vocal chords (all people on Earth shouting together at the top of their voices would develop no more than 10 horsepower) is sufficient to make a sound wave travel a considerable distance before it is damped by the medium.

An investigation of the properties of elastic bodies, however, would lead us too far astray. Note only that waves of both types — longitudinal and transverse — can propagate through solid elastic bodies. Only longitudinal waves can propagate through gases.

Probably the most striking properties of wave motion, its "identification tag", are the phenomena of interference and diffraction. Essentially both are very simple, though diffraction is usually harder to visualise than interference.

Diffraction is due to the ability of a wave to skirt an obstacle. If waves spreading over a pool of water encounter a stone they do not

A few words about the most important properties of wave motion.

produce a sharply defined wave "shade". The waves overlap the "shaded" region.

All other conditions being equal, the "skirting" ability of a wave is the greater the greater the ratio of the wavelength to the size of the obstacle.

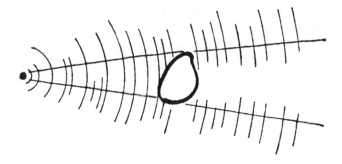

That is why, when a person covers his mouth with his hand, we can hear him speaking, though we can't see his mouth. Sound waves easily bend around his hand, but the diffraction of light waves is much too small for that. In other words, the shorter the wavelength the more difficult it is to observe diffraction.

The diffraction of visible light can be observed by means of a fairly simple set-up; to observe diffraction of X-rays, obstacles of atomic size are required.

Interference is a direct consequence of the phenomenon of superposition. You should not be dismayed by the learned ring of the sentence. The principle of superposition is very clear: If one point is subjected to several actions, we must sum them up to obtain their net result. A sum of even two identical waves may produce different results depending on the phase shift. Incidentally, a suitable analogy to interference is the rule of force composition. Thus, it is theoretically quite possible for utter silence to set in with several people talking loudly in a room. For several reasons this is never observed, but experimental conditions for observing interference can be easily created in optics, acoustics and in studying elastic vibration in solids.

Probably the most difficult property is that of polarisation, but we shall not consider it for a while.

Physicists studying light observed effects which clearly indicated its wave nature. Interference and diffraction, you will recall, were first observed by Grimaldi. After him Hooke, Newton and Huygens all encountered the phenomena in their experiments.

Winding up our cursory remarks on waves and returning (attention!) to the ether.

But if light is assumed to propagate by wave motion, then *there must be a medium in which this motion takes place. This medium is the ether.*

The ether, whose existence Descartes postulated by purely speculative reasoning; the ether, which Newton accepted, though he could never reconcile himself to it completely and preferred to mention as rarely as possible till the end of his days; the ether, for which Huygens finally developed what looked like a reasonable explanation.

What then was the ether of pre-twentieth century physicists?

Investigation of sound waves in air and elastic bodies led to the conclusion that wave motion is possible only in a continuous medium, or continuum. Hence, if light propagates in waves, then obviously the space about us must be filled with a continuum possessing certain wonderful properties.

These properties are all the more wonderful as there is no physical test, save the propagation of light, which provides a means of detecting it.

On the other hand, physicists could not imagine wave motion without some material medium.* Their experience with waves in air, on the surface of water and in elastic bodies forced them to the conclusion that *waves can spread only through a medium consisting of interconnected particles.*

In general, scientists like to reason with analogies, but this was not the case of a more or less suitable analogy: it was an analogy that simply could not be evaded! Hence, the triumph of the wave theory meant a simultaneous triumph for the ether hypothesis. Descartes' idea of the existence of an ethereal *materia subtilis* pervading the whole of space gradually won over the minds of physicists together with the wave theory of light.

The subsequent years were devoted to controversies over the properties of the ether. Huygens' ether was unlike Descartes' ether, and nineteenth century ether was like neither. But, in one form or another, the ether hypothesis took firm root in physics.

If the whole of the universe is filled with this fluid matter called ether, then the problem of an absolute frame of reference possessing some unique qualities is solved. This reference frame is the ever-calm ether, and motion with respect to the ether is absolute motion.

---

*Material here means consisting of minute particles.

The fact that in investigating mechanical phenomena it is impossible to distinguish absolute motion from relative motion is immaterial. We can find light phenomena which will make it possible to observe the calm ether. Thus we shall establish an absolute frame.

Incidentally, if we accept the existence of the ether, it is very natural to assume that centrifugal forces are developed by rotation with respect to the ether. New prospects open up for explaining the nature of inertial and non-inertial reference frames. Interaction through ether makes it possible to explain the mechanism of gravitation. The ether hypothesis is very attractive, even if we did not have to explain the properties of light.

Thus, towards the close of the seventeenth century the knot was tied in physics, which Albert Einstein was destined to cut in the year 1905.

CHAPTER VIII,

which is devoted to the wave theory of light. The patient reader may derive some satisfaction out of an acquaintance with some very subtle and far-reaching conclusions developed from an investigation of the strange effect of double refraction

THE ETHER (continued)

...There is no evidence for its existence, therefore it ought to be rejected.

NEWTON

In the last chapter an essential inaccuracy was made due to the schematic nature of the discourse. It was stated that Huygens was the first to enunciate the wave nature of light, while Newton advanced the opposing corpuscular theory.

This is not quite so. In his works Huygens failed to explain many basic phenomena which are observed in the study of light. For example, he did not realise the periodic character of light, and consequently was unable to offer a satisfactory explanation of interference. He said nothing of diffraction and for all we know

A soliloquy on the ether which gives us additional cause to admire Newton.

162

may not even have been aware of the existence of the phenomenon. Finally, Huygens' theory failed to explain such fundamental properties of light as rectilinear propagation and colour.

In Huygens' and Newton's time the wave theory of light provided a very inadequate description of the effects observed, and it is not surprising that Newton rejected it, even though he was not a dyed-in-the-wool supporter of the corpuscular theory.

To the creator of mechanics and optics a hypothesis was always a hypothesis and he relegated such things to second-class physics. On the other hand, when it was impossible to proceed without a hypothesis, Newton could conjure them up better than any of his contemporaries.

He understood the properties of light better than Huygens or any other physicist of his time, but he was also as suspicious and ironical of his own theories as of the conjectures of others.

As L. I. Mandelshtam remarks, the only reason why Newton did not develop a wave theory of light was that he saw its shortcomings better than Huygens. True, neither was he satisfied with the corpuscular theory. He probably equally "disliked" the ether, though in the course of his life he put forward several clever ether hypotheses.

There is only one way of offering an idea of the remarkable flexibility and inventiveness of his mind, and that is by setting forth his hypotheses. This, unfortunately, we are unable to do here. As a curious fact it could be mentioned that Mendeleyev at the close of the nineteenth century (!) discussed the ether in much the

Though they attached different properties to the ether, Mendeleyev's ether was the twin of Newton's.

same plane as Newton. Moreover, Mendeleyev was so sure of the existence of the ether that he even reserved a zero place in his periodic table for a chemical element "newtonium", that is, ether. That Newton's ether could attract such a scientist as Mendeleyev two hundred years later is in itself an excellent recommendation.

But here is what Newton himself thought of that section of his work. His famous letter to Robert Boyle on the ether ends with an unexpected note (unexpected, that is, if you don't know Newton):

"For my own part, I have so little fancy to things of this nature, that, had not your encouragement moved me to it, I should never, I think, have thus far set pen to paper about them."

This was written after several pages which sparkle with such a firework of clever and subtle hypotheses that in truth the ideas contained in that letter alone would have been sufficient to immortalise Newton's name in the annals of science.

Of course, there may be different ideas concerning Newton's attitude towards hypotheses as a method of scientific research. In this case, however, his ironical attitude towards his own findings prompted by relentless exactingness arouse feelings of profound respect for him as a scientist and a man.

A brief history of the wave theory of light.

Newton's scepticism was forgotten by later generations, and with good reason. By the beginning of the nineteenth century, the wave theory of light had triumphed completely. Up till the very end of the eighteenth century most physicists regarded light as a flux or corpuscules, not waves — partly under the influence of Newton's backing, but mainly due to the shortcomings of the wave theory. The principal proponents of the corpuscular theory were the members of the advanced, powerful school brought together by the French Academy.

The first telling blow at the corpuscular theory was dealt in 1801 by Thomas Young, who offered an explanation of the interference of light on the basis of the wave nature of light. Young's wave scheme provided a sound theoretical explanation of the colours of thin plates and the so-called Newton's rings.* This, however, failed to convince the adherents of the corpuscular theory, who could point to achievements of their own. Laplace, for instance, provided a satisfactory explanation of double refraction by means of the corpuscular theory, and the issue remained open.

In 1818, the French Academy offered a premium for a rational theoretical explanation of diffraction. The idea was that the winning work would prove a corpuscular nature of light. Alas, expectations betray us all too frequently, and the winning work explained all the phenomena of diffraction by the wave theory. Its author was a young physicist, Augustin Jean Fresnel, and he may be regarded the founder of the wave theory of light.

Mention is again made of double refraction without, however, explaining what it is.

The members of the jury, all adherents of the corpuscular theory, accepted the work grudgingly. They were real scientists, however, and Fresnel's work received an excellent review.

---

*Newton's rings represent one of the best proofs of the wave nature of light, and Newton was the first to realise this.

A very edifying episode took place during the debate. Simeon Poisson, celebrated nineteenth-century French mathematician, refused to understand or accept Fresnel's ideas. He pointed out an "absurd" conclusion arising from the wave theory: an opaque shield of a certain size should produce a bright spot in the centre of the shadow.

A most didactic story. It would be difficult to cite a better example of "genuine" misunderstanding.

A test actually produced a bright spot in the very point indicated by theory. Poisson's refutation turned into a brilliant proof of the wave nature of light.

After Fresnel no one questioned the wave theory and neither were there any more doubts as to the existence of the ether.

Yet the ether continued to be a major stumbling block of physics. Moreover, Fresnel's works, while bearing out the existence of the ether, on the other hand served to complicate the situation still more.

The first question that arose when the ether hypothesis was advanced was: "Why don't the planets and all other bodies encounter any resistance in moving through the ether?" The only satisfactory explanation was to regard the ether as an extremely tenuous gas consisting of infinitesimal particles. In this case it could be claimed that the drag of ether was so small as to be experimentally undetectable.

But if the ether was a gaseous substance, according to Newton "a certain most subtle spirit which pervades and lies hid in all gross bodies", then only longitudinal waves could develop in it. Transverse waves cannot form in gases, hence light waves must be longitudinal.

Fresnel, however, demonstrated that light propagated in transverse waves. You might well ask how one could establish whether light propagates in longitudinal or transverse waves if there was no way of observing the particles of the medium — the ether — through which it moves. After all, no one could see what was actually oscillating in a light wave.

Nevertheless, the transverse nature of ethereal vibrations was established beyond doubt. This was proved very elegantly on the basis of symmetry.

Consider a ray of light. By the wave theory, it owes its existence to vibrations of the ether. The vibrating portion of the ether represents a very narrow "tube" whose centre line is the axis of the light beam.

Now, if we assume that the vibrations of the ether in the tube are longitudinal (that is, directed along the axis), and if the properties of the ether are uniform in all directions (isotropy of the ether), then we may assert that

Difficulties of the ether hypothesis.

The first trouble.

Longitudinal or transverse light waves?

the physical properties of a beam of light must possess axial symmetry. This should be elucidated.

Isotropy: an important physical concept.

What is meant by "isotropy of the ether"? Only that its physical properties do not depend on direction. This was demonstrated by experiment. For instance, when a light flashes, the light waves travel absolutely identically in all directions. Similarly, a skier on a vast field of virgin snow has no preference of direction as far as his skiing is concerned. But if he finds a track crossing the field, or if he is on a mountain slope, then the isotropy is disturbed and he can choose a more or less favourable direction.

Axial symmetry of all properties of a beam means that in a light beam all directions perpendicular to the axis are identical. If we turn a light beam about its axis, nothing changes and the physical picture remains as it was. Ordinary light beams possess such symmetry, and there is no experiment to show that any direction perpendicular to the axis possesses some special properties.

All this, it would seem, supports the hypothesis of the longitudinal vibration of the ether. But let us not hurry with conclusions.

Back in 1670, Huygens found that a beam of light passing through a crystal of Iceland spar

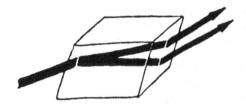

splits into two. A most surprising fact. True, Huygens provided a formal (mathematical) explanation of double refraction, but he failed to understand the physical meaning of the phenomenon. In fact, the best he could do was advance a hypothesis of two ethers (sic!) in Iceland spar.

Newton, naturally enough, strongly criticised Huygens for this conjecture. Sir Isaac had his hands full with one ether and he refused to tolerate more.

As an aside it should be noted that, usually, the more artificial a proposition, the more involved a hypothesis and the less chance of its ever being confirmed. This, of course, is not so surprising. When a hypothesis is false, contradictory facts accumulate rapidly and more and more artificial propositions have to be added to bolster it.

The troubles of double refraction were not over. If we pass a ray of light through two crystals of Iceland spar there will be four beams emerging from the output end. Now if we rotate the right-hand crystal with respect to the other the brightness of the four spots on the screen will change. In some positions of the two crystals with respect to one another we will have only two beams, the other two disappearing! This means that the beams

We finally come to double refraction, which proves the transverse oscillation of light waves.

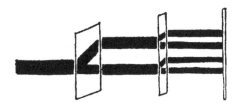

emerging from the first crystal no longer split in the second.[†]

We thus find that the intensity of each of the four beams changes depending only on the relative positions of the crystals. The rays emerging from the left-hand crystal react to the position of the second. In other words, they are "prepared" for their passage through the right-hand crystal. All this is very strange. To all appearances the rotation of the right-hand crystal alters nothing as both are homogeneous and of the same thickness. All that changes is the relative position of their crystallographic lattices. Furthermore, if we turn both crystals together, without altering their relative position, we find that the intensity of the four emerging beams is constant and does not depend on the rotation.

This immediately suggests that the effect is due to some property of the light beams emerging from the left-hand crystal. This light differs in some way from ordinary light, but how?

For over a century the "Huygens effect" remained without an explanation. This is hardly surprising, for if we say that light is transmitted by longitudinal waves through the ether, which is itself isotropic, then it is impossible to comprehend how two beams of white light of equal intensity can differ from each other.

[†]Two beams are obtained at the output when the respective crystallographic axes are either coincident or at right angles. Therefore, when we turn the right-hand crystal through a full angle of 360° we observe two beams instead of four in four positions of the crystals.

But as soon as we assume that light represents transverse oscillations we immediately obtain a new parameter: the direction of oscillation of the ether particles. If the vibrations are transverse, then in the tube cut out of the ether by the light beam we have a longitudinal plane, namely that in which the ether particles are oscillating.

The guiding principle for explaining double refraction: transverse light waves.

Why then can't we observe such a plane in an ordinary light beam? This is quite simple. Recall that white light is a mixture of light waves of different length, a mixture of different colours (Newton again!). We may well assume that in an ordinary light beam we also have a uniform mixture of light waves undulating in different planes, which is why there is no isolated plane.

But what if Iceland spar is capable of sorting out light rays in some way? The two beams emerging from the first crystal are characterised by different planes of oscillation. This hypothesis may not seem very convincing, all the more so as the picture is further complicated by the emergence of four beams. Let us try a similar experiment with a crystal of tourmaline.

We take a properly cut lamina of tourmaline and direct a beam of light perpendicular to its surface. It passes through the lamina without

An elegant experiment with a crystal of tourmaline.

refracting.[‡] Neither does its intensity change if the lamina is revolved about the axis of the beam.

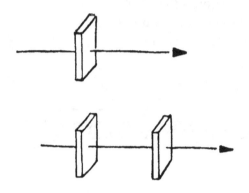

Now let us complicate the experiment by passing the ray through two laminas. If we revolve the second lamina about the axis of the beam, we see that the intensity of the light beam increases to a maximum and then gradually fades to nill and no light passes through. The phenomenon repeats itself on further rotation.

Conclusion.

We see that the beam of light passing through the two laminas is sensitive to their relative position. Assuming light waves to be longitudinal, it is impossible to explain how and why the rotation of the laminas should affect the beam. Hence, we must accept that the ether in the beam of light oscillates transversely. The beam emerging from a tourmaline crystal is polarised; which brings us to the definition of a polarised wave.

[‡]Tourmaline also possesses the property of double refraction, but one of the beams is completely absorbed in the crystal.

A transverse wave is said to be polarised if the particles of the medium through which it is passing are oscillating in one plane.

Thus, light emerging from crystals of Iceland spar and tourmaline is plane polarised (probably!).

Was it worth the time dwelling at such length on these experiments, especially Huygens' experiments with Iceland spar? After all, tourmaline alone establishes the transverse nature of light waves in a simpler way.

It was. We often hear that seventeenth-century scholars had a much easier life. All they had to do was stage a simple experiment or hazard a commonplace conclusion — and a new word in science was said. I think, therefore, that it would be useful to give a more careful analysis of the "simple" conclusions of Galileo, Newton and Huygens.

An edifying discourse on the complexity of science and the possibility of other attempts to explain double refraction.

We are obviously in no position to duly examine the problems of the past, but I should like you to realise how surprising and inexplicable new effects always seem, regardless of the age in which they are observed.

Strictly speaking, even the experiment with tourmaline is not sufficient to conclude that the ether in a light wave oscillates only transversely. So far not a single fact has been cited which could disclaim the existence of longitudinal waves. More, we might even advance a new hypothesis, say, that the ether particles represent tiny randomly spaced "light magnets". These particles could be oscillating longitudinally in a light beam. In an ordinary light beam all the randomly orientated little ether magnets are vibrating. On passing through a tourmaline crystal for some reason or other only specifically orientated little magnets start oscillating longitudinally. Then, though the vibrations are longitudinal, a favoured direction appears.

The easiest thing would be to say that all this is nonsense. But there is a difference between saying and proving.

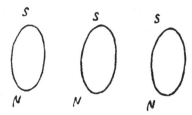

We could also attempt to explain light polarisation by the corpuscular theory. All that is needed is to introduce a hypothesis that the light corpuscles are like little magnets and that a crystal arranges them in some specific order. Incidentally, the author of this hypothesis is none other than Sir Isaac Newton. He was the first to realise the tremendous importance of Huygens' experiments in which the "connate properties" of light, to use his expression, displayed themselves.

Transverse light waves and the ether hypothesis.

In short, we should not be surprised that many physicists were reluctant to accept the hypothesis of the transverse nature of light waves. It seemed most unnatural to them, for to concede that the ether transmitted transverse waves meant rejecting the gas-model of the ether! Gas cannot transmit transverse waves! It was necessary to reappraise old concepts and give a solid-state analogue of the ether. But how then to explain the motion of heavenly bodies through it without friction? And that was not all, for solids and liquids can transmit both transverse and longitudinal waves. But Fresnel's and Arago's experiments[§] in the early nineteenth century could be explained only on the assumption that there were no longitudinal oscillations in light waves. This was just too bad!

In mechanics it was proved that when an absolutely transverse wave hits an interface of two elastic media the reflected and refracted waves must contain a longitudinal component. This is not the case in the ether as reflected light consists only of absolutely transverse waves!

A satisfactory hypothesis was hardly advanced to explain this when physicists were confronted with more baffling discoveries.

Michael Faraday found that the plane of light polarisation rotates under the action of a magnetic field. It appeared that light and electromagnetic phenomena were closely related and the "luminiferous" ether was at least akin to the "electrical" ether!

[§]We shall not examine these works.

CHAPTER IX,

a perusal of which may help the reader to form a slightly better idea of how "simple" it is to study physics

THE BIRTH OF THE STATIONARY ETHER

...There is no evidence for its existence, therefore it ought to be rejected.

NEWTON

This chapter brings us, so to say, to the foothills of Einstein's theory. Everything that follows is devoted essentially to one question: "Is it possible by any experimental means to detect the stationary ether, and hence to produce an absolute reference frame?"

Enviable fate of the ether.

The attitude of nineteenth-century physicists towards the ether hypothesis was much like that of parents towards an only child. They forgave the ether everything: its more than strange properties of a super solid (transverse oscillation of light waves) and at the same time its exceptional tenuity which followed from the absence of any effect on stellar or planetary motion, and its unconventional behaviour in some solids (two ethers in Iceland spar). Later the premise of two ethers was supplanted by that of varying elasticity of the ether along different crystallographic directions (Neumann, 1835), but this could hardly be regarded as a happy solution.

Physicists reconciled themselves to everything because without the ether, without some medium, it was impossible to imagine how electromagnetic waves travelled through space.

In our time we speak freely of space itself being capable of transmitting electromagnetic and gravitational waves without associating this ability with the presence of a medium permeating the whole of the universe.

Mechanical models of the ether have been rejected in favour of the new field concept. Without going into subtleties we will just note that contemporary physics has given up attempts to represent electromagnetic waves by analogy with waves in mechanical media or gases. At this juncture we shall merely state the fact that some waves possessing certain specific properties are capable of propagating through space.

As yet, naturally, we refrain from expounding the contemporary view.

We know today that the hypothesis according to which space is permeated with a medium with properties analogous to those of gases or elastic bodies — the ether, that is — is insolvent and disproved by experience. In short, in the question of the ether the physicists have returned to the method of principles. But we can imagine how difficult it was to give up the very tangible concept of an intangible ether as an elastic medium permeating all of space.

A brief philological remark is called for. When we say that the theory of relativity has banished the ether from physics all we mean is that the idea was rejected of the whole of space being pervaded by a medium consisting of particles. We only say that waves can travel through space. You can call this space by the name of ether if you like. This is merely a question of nomenclature.

The classical ether passed away when it was established that for optical phenomena, as in mechanics, there was no exclusive reference frame.

A sentimental introduction.

It took more than two hundred years of research before this was proved and Einstein formulated his theory. Hundreds of experiments, scores of theories and the talent and effort of generations of physicists paved the way for Einstein's triumph. Each contributed his share according to his ability: those whose works were forgotten almost as soon as they appeared as well as those whose names have gone down in the annals of science.

There is probably no more exciting story in science than that of the development of the ether theory. Several times it seemed that all doubts had been wiped away. Yet a decade or so would pass and new experiments would shatter theories which had seemed so sound a few years ago.

Regretfully, we are unable here to offer even a cursory outline of that great and difficult road. We shall limit ourselves to two works made at the dawn of research into the phenomena of light. They have been selected not so much because of the important part they played in the story of the ether and the science of light as for the fact that by tracing the remarkable, unexpected and very bold conclusions of their authors (both of them otherwise undistinguished scientists) we may gain an understanding of physics.

The first work.

In 1676, Olaf Roemer, a Danish mathematician and astronomer, observed strange irregularities in the motion of Jupiter's closest satellite: its eclipses regularly failed to coincide with the predicted times. This was most puzzling.

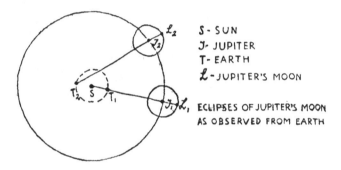

S - SUN
J - JUPITER
T - EARTH
$\mathcal{L}$ - JUPITER'S MOON

ECLIPSES OF JUPITER'S MOON
AS OBSERVED FROM EARTH

Fact number one.

The orbital period of the satellite is constant: observations carried out at different times of the year invariably gave 42 hours 47 minutes 33 seconds. There were some variations, but they were well within the accuracy of the measurements, and the mean value was constant, which, of course, was quite natural.

Mystification! The apparent orbital period of Jupiter's moon is variable. But this could not be observed in Roemer's time due to the low accuracy of instruments.

Furthermore, the orbits of Earth and Jupiter and the velocities of both planets were well-known to astronomers. Obviously, then, if one knew the time of any eclipse of the satellite, one could easily forecast the beginning of any subsequent eclipse. All that was necessary was some painstaking but essentially simple calculations.

Consider: given, three moving bodies; to determine the intervals at which they will be aligned. Let the eclipse be observed when Jupiter, the Earth and the satellite are in position 1 (see drawing). Knowing the orbital period of the satellite, we can calculate the time of all other eclipses in the course of a year. But instead of this simple picture astronomers came up against puzzling fact number two.

Fact number two.

The eclipses were found to occur later than predicted. For six months this time-lag increased till it became several minutes. Then, for the next six months, the time-lag decreased until it disappeared completely.

For half a year the satellite seemed to be moving a bit slower, and in the next half a year a bit faster, in its orbit than as observed in position 1. The inferrence was that the satellite was perturbed by some unknown force which accelerated or retarded it, the action having a period of about one terrestrial year.

What could be the cause of this perturbation?

Roemer advanced the bold hypothesis that the time difference was due not to the rotation of the satellite, which was uniform, but to the fact that *the velocity of light is finite, with the result that an observer on Earth sees the apparent orbital period of the satellite to change.*

Remember that, as far as we know, Roemer's instruments were not accurate enough to detect the time-lag of *each* successive orbit of the satellite around Jupiter. But the lag of fractions of a second gradually accumulated till the difference between the predicted and observed time of the eclipses reached several minutes.

Incidentally, something analogous to the lagging of Jupiter's moon is observed in everyday life. Often someone says: "This is a very accurate watch: it lags only one minute in a month." The accuracy of an ordinary wristwatch is insufficient to detect a lag of a few seconds a day. But the seconds gradually build up into the perceptible lag of one minute a month. We might see in a few days that the second hand is slightly behind time, though when we check the watch by the radio time signal we can't detect a lag of one second and we say that the watch is correct. This is probably clear, but with Jupiter's moon matters were further complicated by a number of side effects.

Cassini, the celebrated French astronomer, under whom Roemer worked, first accepted his theory. Later Cassini withdrew his support because the apparent motions of the other Jovian satellites seemed to contradict Roemer's conclusions. As was often the case, Roemer never lived to see his theory completely recognised.

The explanation of the apparent nonuniformity of the rotation due to the finite velocity of light is a simple one. Consider two positions of the Earth and Jupiter for which we measure the time between successive eclipses. Observe that in configuration 1 the distance between the Earth and Jupiter is gradually decreasing while in configuration 2 it is increasing. Now, assuming the velocity of

An attempt to explain popularly the apparent discrepancy in Roemer's observations.

light to be finite, let Jupiter, its moon and the Earth be in configuration 1, and let the satellite have been eclipsed by Jupiter at time t_1. The light waves reaching Earth at that instant were emitted from the surface of the satellite at some earlier moment. In other words, we see the image of the satellite in a point through which it has already passed.

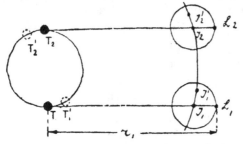

Probably the first exact analogy.

Similarly, we fail to see a fast-flying aircraft if we look for it in the quarter of the sky from which we hear the roar of its engines: by the time the sound reaches us the plane is already far away.

As is often the case, equations help to get to the root of the question.

The image of the satellite disappearing behind Jupiter's face reaches us not at time t_1 but after a time interval Δt_1 which it takes for light to travel the distance r_1 from Jupiter's satellite to the Earth. This interval is equal to

$$\Delta t_1 = \frac{r_1}{c},$$

where c is the velocity of light.

A terrestrial observer will record that the eclipse took place, according to his watch, at time

$$t_t^{\text{obs}} = t_1 + \frac{r_1}{c}.^*$$

---

*An observer naturally records time t_1^{obs}. Time t_1 can be established only by calculating it if r_1 and c are known.

The same thing happens when the following eclipse occurs (in about two days), and the astronomer enters in his log that the eclipse began at time $\tilde{t}_t^{obs} = \tilde{t}_1 + \dfrac{\tilde{r}_1}{c}$, where $\tilde{r}_1$ is the distance from the Earth to Jupiter at the time $\tilde{t}_1$ of the second eclipse. The time interval between the beginning of the two eclipses is

$$\Delta t_t^{obs} = \tilde{t}_1 - t_1 + \frac{1}{c}\left(\tilde{r}_1 - r_1\right).$$

But, as you will recall, in configuration 1 the distance between the Earth and Jupiter is steadily decreasing, hence $\tilde{r}_1 < r_1$ and the second term in the right-hand part of the equation is negative. True, the velocity of light c is very great and the second term is therefore very small as compared with the first, but nevertheless the measured time interval is slightly smaller than the actual time lapse between the two eclipses. The foregoing may be repeated with respect to configuration 2, and we obtain:

$$\Delta t_2^{obs} = \tilde{t}_2 - t_2 + \frac{1}{c}\left(\tilde{r}_2 - r_2\right).$$

There is one important difference, however, for in configuration 2 the distance between the Earth and Jupiter is increasing, and $\tilde{r}_2 > r_2$. Hence, the second term in the equation is positive and the time interval Δt_2^{obs} is slightly greater than the actual time lapse between the eclipses (and, obviously, $\Delta t_2^{obs} > \Delta t_1^{obs}$).

Knowing the motions of the Earth and Jupiter, we can determine the distance between them at any given time. With these data we can, by simple computations, readily find the velocity of light.

Roemer's calculations were far from accurate, and he obtained approximately 215,000 km/sec as the velocity of light (actually it is 299,976 km/sec).

Our examination of Roemer's method was a bit more detailed than usual. But even so we have considered only one difficulty: the contradiction between the apparent uniformity of rotation of the satellite and the long-term forecasts of its eclipses. Roemer had to overcome equally difficult obstacles. It was not sufficient to connect the fact of errors in long-term forecasts of eclipses with the finiteness of the velocity of light. It was necessary to process a vast amount of assorted experimental data so contradictory that Cassini later rejected Roemer's theory.

More edifying remarks.

Everything seems so simple when all is said and done and all doubts are removed. This is especially so when acquaintance with the facts is superficial. A closer scrutiny, however, reveals the difficult path progressed before such apparent simplicity could be achieved, the volume of research and the doubts that had to be tackled and overcome. The hackneyed aphorism that "the greater an idea the simpler it is" hardly agrees with the facts. It would be more appropriate to say: "The better an idea is understood by others the simpler it seems", though sometimes it is rather the less we understand a thing the simpler it seems.

Now let us have a look at the second work which played such an exceptional part in the theory of light and ether. lncidentally, it was, in part, the work of chance.

Ever since the Copernican system was postulated, its adherents sought to prove the Earth's rotation by detecting the apparent annual displacement of the fixed stars, the so-called parallax.

A very gradual approach to the story of aberration of light, an effect remarkable both for its physical nature and its history.

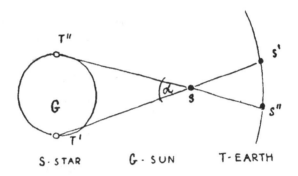

S · STAR   G · SUN   T · EARTH

α THE ANGLE CALLED THE ANNUAL
PARALLAX OF THE STAR

The idea is simple enough. When the Earth is in position T', a star is seen in point S'. Half a year later, from position T'', we see the star at S'', and in the course of a year it will follow the path $S' \rightarrow S'' \rightarrow S'$.[†] In other words, the apparent motion of the star requires that our telescope be directed at different angles in order to see it, that is, to different points of the sky.

[†]For the sake of simplicity the case of a star lying in the plane of the Earth's orbit (the plane of the ecliptic) is considered.

There is a fine point which is of no consequence to us. A star not in the plane of the ecliptic is seen to move in an ellipse, just as the Earth's orbit would be seen from the star.

As the distance to the closest star is many times greater than the diameter of the Earth's orbit, the annual parallax is very small, which is why sixteenth-century astronomers were unable to observe it. In fact, the largest parallax, that of Proxima Centauri is equal to 0".75. This is the angle subtended by a hair at a distance of 60 feet![‡]

Some interesting facts.

The famous Danish astronomer Tycho Brahe attempted in vain to determine the annual parallax of the Pole Star. When he failed he turned into a violent opponent of the Copernican system.

In the seventeenth century the accuracy of astronomical observations improved considerably, and sure enough a small displacement of stars was actually observed. It was attributed to the annual parallax, which meant further confirmation of the Copernican system. But then James Bradley, studying the annual displacements of many stars, came to the conclusion that they had nothing to do with the parallax as they did not coincide with theoretical findings. They simply had nothing in common.

Firstly, all stars without exception lying in the plane of the ecliptic were observed to move twice through an arc of 40".9 in the course of a year. Furthermore, all the stars not lying in the ecliptic plane described ellipses the larger axis of which all measured 40".9.

If these displacements were to be attributed to the parallax then the fantastic assumption

---

[‡]Incidentally, the annual parallax of a star is used to measure its distance from the Earth. Today the parallax is determined with an accuracy of 0".01. This is the angle subtended by a human hair at a mile away!

had to be made that all stars lay at the same distance from Earth. However, even such a desperate move would fail to save the situation. The motion discovered by Bradley displayed features which could not be explained if it were treated as parallactic displacement. The thing is that, if the observed displacement was due to the parallax, then in the two positions when the Earth, Sun and star are aligned, the star should be observed at one and the same point in the sky. Bradley, however, found that it was precisely in those two points that the star was most displaced in respect to its equilibrium position in the sky.

Bradley found a bold and ingenious explanation of the phenomenon. Let us assume, he says, that the velocity of light is finite. Light is a stream of minute corpuscles reaching the Earth from a star (Bradley was a firm supporter of the corpuscular theory of light). Now, since the Earth travels in its orbit with considerable speed, the apparent picture of the stellar sphere must differ from its actual looks. Bradley's idea is easily explained.

Another reasonably
close analogy.

Suppose at some observatory an astronomer has pointed his telescope straight at the zenith, that is, perpendicular to the ground. To facilitate our reasoning, let the telescope be a reflector, the upper end of which is open. Suppose now that it starts raining and the raindrops are falling straight down. If the telescope is left in the open, the rain will patter uniformly on the mirror. A raindrop falling along the optical axis will hit the mirror square in the centre.

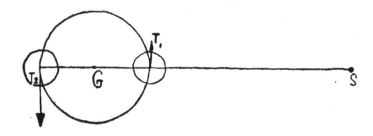

Now suppose that the observatory, telescope and absent-minded astronomer are all on board a fast-sailing vessel, and again an absolutely vertical downpour begins. The picture changes. By the time a drop falls down the telescope the latter is "carried away" from under it and the drop does not fall along the centre line. It appears to be slanting in the opposite direction of ship's motion. As a result, more rain falls on the left edge of the mirror than on the right (see drawing on next page).

For the raindrops to fall parallel to the centre line of the tube, it must be tilted at some angle to the right. If the astronomer took it into his head to determine the direction in which the raindrops were falling according to the inclination of his telescope, he would be grossly mistaken.

Now back to the stars. The image of a star in the zenith is projected on the mirror of our telescope. If the Earth were at rest, the stream of light corpuscles coming from the star would be exactly parallel to the centre line of the tube and would reach the eyepiece.

But what if the Earth is moving? Then, by the time a light corpuscle falls the length of the tube, the latter is carried away and the light beam slants away from the eyepiece. To make the corpuscles fall along the centre line the tube must be tilted.

The angle of inclination is easily determined. If the velocity of the light corpuscles is c and the velocity of the telescope (i.e., of the Earth) is v, then

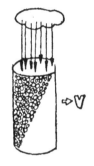

$$\tan \varphi = \frac{v}{c}.$$

An astronomer determines the direction towards a star by the direction of the centre line of his telescope at the instant when the image of the star is projected in the centre of the visible field (on the optical axis). According to Newton's corpuscular theory, the image of an object is projected along the optical axis of the tube only if the light corpuscles are flying parallel to the centre line. And, as we have just seen, due to the motion of the Earth the corpuscles can travel along the centre line only when the tube is pointed not directly at their source (the star) but slightly away. This is the phenomenon of light aberration.[§]

[§]Aberration means literally, deviation. It is not surprising, therefore, that the term is used to denote physical

Incidentally, probably everyone has had occasion to observe the "aberration" of a vertical downpour. Judging only by the paths traced by raindrops on the lateral window pane of a moving vehicle, the rain would seem to be slanting.

An important remark.

It should be pointed out that if the Earth were moving in a straight line and with constant speed with respect to the fixed stars, we would have no way of establishing the existence of aberration of light. In all such experiments telescopes would be tilted at the same angle with respect to the true direction to a star, no aberrational movement of the star across the sky would be observed and the existence of aberrational displacement could only be conjectured by calculation. As the drawing shows, the apparent displacement of stars is due to the fact that the Earth's velocity is differently directed at different points of its orbit.

Bradley's theory provided an excellent explanation for the apparent displacement of stars. In particular, it became clear why the maximum angular displacement of all stars is equal, for it depends solely on the ratio of the orbital velocity of the Earth to the velocity of light.

Incidentally, the magnitude of the angular displacement could be used to determine the velocity of light, which Bradley did and found it to be 303,000 km/sec, an accuracy of one percent.

phenomena of very different nature which cause distortions of light rays. We know of "chromatic", "spherical", "longitudinal" and several other aberrations.

The aberrational displacement of stars served, furthermore, as an excellent proof of the validity of the Copernican system. In short, Bradley discovered a much more interesting phenomenon than he had expected.

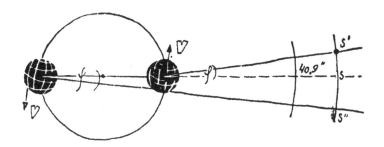

As to parallactic displacement, it was discovered only in the middle of the nineteenth century, as it was too small to be detected by eighteenth-century instruments.

Science had triumphed.

All this is very well, but we have forgotten that the corpuscular theory of light was found to be wrong! Therefore we cannot be satisfied with Bradley's explanation of aberration. We must explain the phenomenon from the point of view of the wave theory, otherwise it hangs in mid-air.

An explanation of aberration from the wave point of view was offered by Thomas Young in 1804. The problem proved to be much more rancorous than originally suspected.

Young assumed that the ether was not carried along by the motion of the Earth, that the Earth was hurtling through an ever-calm ether sea and its velocity with respect to the ether particles was equal to its orbital

The stationary ether! Attention!

velocity of about 30 kilometres per second. Only in this case, in the case of an absolutely stationary ether, could the aberration of light as computed in the wave theory coincide with the value predicted by the corpuscular theory and actually observed.

We shall now proceed with a brief outline of Young's reasoning, but there is first one point to be made. In their attempts to reconcile the aberration of light with the wave theory, physicists for the first time came up against the crucial problem of the ether theory which ultimately spelt its downfall.

How does the ether act on the moving Earth? How does the Earth's motion with respect to the ether affect optical and electromagnetic phenomena? Can motion relative to the ether be detected experimentally? All of which boiled down to the ultimate question: *Does there exist an absolute frame of reference — the stationary ether?*

Thus, the problem of aberration in the wave theory of light became an important question of principle. Its solution is pretty involved so we shall restrict ourselves to several general qualitative remarks which, nevertheless, offer a correct idea of the gist of the problem.

The light waves beamed forth by a star spread concentrically away from it through the stationary ether.[¶] Suppose that the moving Earth does not drag the ether along. Then the waves travelling inside a telescope will appear to be deflected to the left of the centre line.

Aberration of light and the stationary ether. A very important passage.

[¶]Remember what we mean by the stationary ether: *The ether is assumed to be at rest in respect to the frame of fixed stars.*

Now, if the Earth carried the ether along, there would be no aberration of light! The fact that aberration is observed indicates that the Earth does not drag the ether along. Therefore, as the Earth is moving relative to the fixed stars, there should be an "ether wind" around it.

The next question, naturally, is: Can the "ether wind" be detected by means of other optical phenomena? Simple theoretical reasoning led immediately to a positive answer.

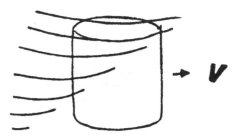

For example, the refractive index of light in the case of the ether not being carried along by the Earth should differ depending on whether the Earth is moving towards the source of light (the star) or away from it. An experiment was carried out, but nothing was observed, though the precision of the instruments was sufficient to detect the predicted effect. This was most disconcerting. On the one hand, aberration of light seemed to support the stationary ether theory, on the other, experiments with the refractive index refuted the theory.

Arago's experiment — 1818!

In addition there were some snags in the aberration effect. The tilt of the telescope is determined by the ratio of the distance it travels in the time it takes for the light to travel down the tube to the length of the tube. In other words, the tilt is given by the ratio $\frac{v}{c}$.[**] An important point to be borne in mind is that c here is the velocity of light *inside the tube*. Accordingly, a most effective experiment was staged with a telescope filled with water.[††] The speed of light in water is about 3/4 of its speed in air. Consequently, the aberration angle in a water-filled telescope should have increased by about 4/3.

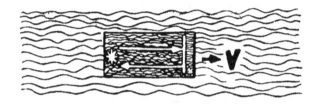

Contrary to all expectations, experiments demonstrated that it remained unchanged. This was most unfortunate. Fresnel, however, was able to overcome the difficulty by advancing an ingenious hypothesis explàining ether drag

---

[**]To be more precise, $\tan \varphi = \frac{v}{c}$. But the ratio of the Earth's velocity to that of light is $\dfrac{30 \text{ km / sec}}{3 \times 10^5 \text{ km / sec}} = 10^{-4}$.

The tangent of such a small angle (measured in radians) is equal to the angle itself to a high degree of accuracy.

[††]The precise measurement was done by the English astronomer Airy, who carried out the experiment in 1871. But if we are to believe Michelson (*Studies in Optics*), the experiment was staged at least before Fizeau's experiments (1851). Michelson, however, gives no specific reference.

by continuous media. He postulated that the ether was denser in continuous media than in vacuum. Therefore, the ether pervading the vacuum of outer space was not carried along by a moving body, whereas the ether permeating a body was partly carried along by it. According to Fresnel, the quantity of ether flowing into a moving body should be equal to the quantity flowing out. But since the density of the ether inside a body is greater than outside, the quantity of inside ether remains constant only if its velocity with respect to the body is less than that of the "external ether". This theory provided a satisfactory explanation of refraction tests and water-filled telescope experiments.

Calculations based on Fresnel's theory confirmed the principle that the index of refraction should in fact change depending on the motion of a body relative to the ether (the *ether-drift*). Similarly, aberration in a water-filled telescope should differ from aberration in an ordinary tube. But in Fresnel's theory, this effect was very small, the relative change in the refractive index being of the order

$$\frac{v^2}{c^2}$$ instead of $\frac{v}{c}$, that is, of the "second order

of smallness" (for the Earth $\frac{v^2}{c^2} \approx 10^{-8}$). The

difference was thus contended to be so small as to defy experimental verification, for of all motions relative to the ether within reach the fastest was that of the Earth (30 km/sec).

Similarly, the aberration observed in a water-filled telescope was supposed to differ

First mention of an effect corresponding to the second power of the ratio $\frac{v}{c}$. The search for it engaged physicists for almost the whole of the nineteenth century.

from the ordinarily observed aberration by a quantity of the order of $\dfrac{v^2}{c^2}$. In those days there were no instruments capable of detecting such infinitesimal changes. Thus, as a result of Fresnel's appeal, the death sentence on the ether was deferred.

We have spent quite some time on the story of the ether. Most instructive in the whole tale is, perhaps, the tenacity with which physicists clung to the idea. Ether theories followed one another in rapid succession: the "vortex-sponge" ether, the "gyro-elastic" ether, the "labile" ether, etc. Then came the concepts of the dragged, partially dragged and stationary ethers! It would be wrong, however, to ridicule offhand all these ether theories, many of which were postulated by outstanding scientists. They were brilliant, subtle and imaginative, and much talent and ingenuity went into their formulation. Their sole purpose was the salvation of the wave theory of light, for physicists could not imagine waves without a medium consisting of particles.

Yet as time passed the ether became more and more of a freak among other physical substances.

For once, no one seemed to be able to formulate a theory capable of providing a satisfactory explanation of the whole complex of available facts.

Secondly, the hypothetical ether had to be endowed with such remarkable qualities, and such strange assumptions had to be made regarding it, that scientists found it difficult to reconcile themselves to such an ether.

The situation with the ether was aptly summed up by Lord Kelvin, who remarked that one might accept or reject such theories, but one could certainly not be satisfied by them. To crown the ether's many troubles, there still remained unsolved the question: *Does motion relative to the ether affect optical phenomena, at least in the "second order of smallness", or not?* If it were found to have no effect, then it would be necessary to discard Fresnel's partial ether drag theory, the last shaky support of the ether.

Here is a summary of the state of affairs at the time.

1. The wave nature of light suggests the existence of the ether, a mysterious material medium. This medium possesses strange and remarkable properties, but it is difficult to visualise the propagation of light waves without it.

2. The aberration of light indicates that the ether is not carried along by the Earth, as aberration could not be observed in a dragged ether.

3. It follows from this that the Earth's motion in respect to the fixed star frame should display itself in many optical phenomena (notably in a change in the index of refraction).

These conclusions, naturally, apply to the state of affairs in the nineteenth century.

Theoretically, the aberration angle in a water-filled telescope should differ from the aberration angle in an ordinary telescope.

4. Experiments disprove item 3. The Earth's motion relative to the ether has no visible effect on light phenomena. All this is disconcerting, but....

5. The day is saved by Fresnel with his partial ether-drag theory. According to Fresnel, the idea expressed in item 2 is correct in principle, but the effect is too small to be detected. Motion relative to the ether displays itself only in relations of the second order.

6. As yet the accuracy of experiments is inadequate to observe an effect to the second order of smallness. The problem, therefore, continues to be outstanding.

CHAPTER X,

the chief merit of which lies in a rather detailed account of the Doppler effect and Michelson's experiment, and the chief fault of which is an abundance of soliloquising. In this chapter the reader finally parts with the ether and is ready for the theory of relativity

RISE AND FALL OF THE STATIONARY ETHER

...There is no evidence for its existence, therefore it ought to be rejected.

NEWTON

And so, the question of an absolute frame of reference — the stationary ether — remained in the air. It might be worth recalling that the fight flared up around the principle of relativity. If the motion of any frame (say the Earth) relative to the ether affects optical phenomena, then Galileo's principle is not valid as far as optics is concerned. If, on the other hand, it has no effect, then it is valid.

According to Fresnel, the motion has no effect "in the first order of the ratio $\frac{v}{c}$". This statement was sometimes called the practical principle of relativity.

Some cursory information on the development of the ether theory in the nineteenth century. The ether and the electromagnetic theory.

Then, in the eighteen sixties, the problem of the ether received a new turn and matters got even more involved.

As mentioned before, Faraday was the first to establish the connection between optical and electromagnetic phenomena. The importance of his work, however, is much greater. He elaborated a sound experimental basis for the further study of electromagnetic phenomena, and after him research in this field began to develop at a tremendous pace. In fact, Faraday in the science of electricity occupies the same place as Galileo in mechanics.

A few words about Maxwell's theory.

Galileo was bound to be followed by Newton. Similarly, in 1865, James Clerk Maxwell formulated a comprehensive theory of electromagnetic phenomena. The resemblance between Maxwell and Newton is due not only to the fact that Maxwell's work can be ranked alongside of the *Principia* and not only to the fact that, like Newton, he developed a harmonious theory of an entirely new class of phenomena.

Maxwell's discovery represented a complete triumph for the method of principles which, furthermore, he employed in an entirely new form.

Speaking most schematically, Maxwell produced an equation which provided a comprehensive description of all known electromagnetic phenomena. But that is not all. Among the solutions of his equations there were such which, it seemed, related to nothing at all. According to them, electromagnetic waves should be capable of propagation through "empty" space. In Maxwell's time such waves were unimaginable. It followed from his equations that, firstly, the waves were transverse and, secondly, that they travelled with finite speed.

What is meant by "transverse electromagnetic waves"? What is it that oscillates in them? It followed from the equations that an electromagnetic wave was produced by the oscillation of vectors of electrical and magnetic fields perpendicular to the direction of the wave.

Now, where did Maxwell get the speed of his waves?

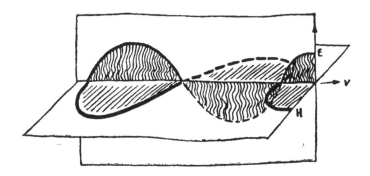

His equations included a constant whose dimension coincided with that of velocity. Maxwell ventured a bold proposition. He assumed that the periodic solutions of his equations in fact described real electromagnetic waves, the unknown constant acquiring definite physical meaning, viz., the velocity of propagation of the waves.

The quantity could be measured purely "electromagnetically". This was done, and it was found to be 310,000 kilometres per second. This you will notice, is very nearly the speed of light. The measurements then were, of course, far from accurate and subsequent experiments gave for the speed of propagation of electromagnetic waves the figure 299,792 kilometres per second.

Maxwell's remarkable idea.

•

Maxwell, however, went further. Not content with postulating the existence of "waves of electrical and magnetic force" which no one had ever heard of before, he used the fact that their hypothetical velocity was so nearly that of light to conclude that light itself was "an electromagnetic disturbance in the form of waves".

This was a very bold and daring idea. Observing that the constant in his equations was of the order of the velocity of light, Maxwell decided: "There must be something in this. Most likely light and my electromagnetic waves are of one and the same nature."

It took time for Maxwell's theory to be recognised, but by the close of the century hardly anyone ventured to challenge its validity.

And what was the fate of the ether with the appearance of the electromagnetic theory? It was hardly enviable. The demands imposed on the ether after Maxwell became more and more exacting, for now it had to explain electromagnetic phenomena as well. On the other hand, experiments with electricity and magnetism provided new opportunities for verifying the ether theory.

A detailed account of the downfall of the ether is, obviously, impossible here. As a result you may not be a hundred percent convinced of the inevitability of Einstein's conclusions.

The author is true to his word and soliloquises volubly. More edifying remarks.

Well, this may not be so bad, after all. You will realise, of course, that the easiest thing would be to outline several decisive experiments and then declare: "As a result of this and that the ether hypothesis ceased to hold water." It could be done very convincingly and the reader would surely believe it, all the more so as it is the truth. But after such explanations the reader usually feels somewhat bewildered. "How is it," he wonders, "that scientists failed to realise this for so many years?" One might even feel superior to, say, Maxwell or Newton.

In my view such thinking is worse than sheer ignorance, which is why throughout this narrative I have sought to steer the reader away from such thoughts.

To tell the truth, most of the problems raised in our discourse have been analysed rather poorly. This is hardly surprising, but it is a point that should be borne in mind. The latter remark is especially applicable to such a general problem as the ether theory. We can discuss its vagaries only superficially, and we must confess that our story offers but a vague and schematic idea of the struggle which raged in the nineteenth century. Furthermore, you need not doubt that any educated physicist of the mid-nineteenth century would easily refute you if you attempted to prove the insolvency of the ether theory merely on the basis of what has been said here.

In the previous chapter we mentioned Fresnel's ingenious theory which sought to explain the behaviour of the ether in continuous media. Fresnel explained why any experiment designed to detect a change in optical properties in continuous media relative to the ether was bound to give deviations only to the "second order" of the ratio $\frac{v}{c}$.

The state of the stationary ether hypothesis on the eve of Michelson's experiment.

But could not an experiment be devised without involving continuous media so that the ether-drift could be observed to the first order of the ratio $\frac{v}{c}$? All such attempts were in vain.

Nature seemed to be laughing at the scientists. All feasible experiments made it possible to observe the ether-drift only to the second order of the ratio $\frac{v}{c}$.

Several ideas of possible experiments "of the first order" were voiced,* but they all proved impracticable due to measurement difficulties. The state of affairs at the time is summed up by Academician Vavilov as follows: "A curious situation arose. Effects of the first order were bound to exist in the stationary ether, but *there was no means of measuring them.*" This situation haunted physicists for about half a century. Meanwhile the ether theory thrived and no decisive argument could be presented against it.

Achievements of the ether theory.

On the contrary, in the interim between Fresnel's hypothesis of the partial drag of the ether in continuous media and Michelson's experiment (an account of which is forthcoming) the stationary ether theory scored some major achievements.

Firstly, it explained the aberration of light at a single stroke.

Secondly, the ether held its own in the face of the accusation that it was impossible to detect any effect of motion relative to it "in the first order". You will recall that Fresnel explained the absence of effects of the "first order" in experiments with continuous media; what is

---

*Maxwell, for example, pointed out that Roemer's experiments could be used to detect the motion of the solar system as a whole relative to the ether to the first order.

more, his theory was brilliantly confirmed in 1851 by Fizeau. Without going into the details of Fizeau's experiment, we may note that none other than Michelson wrote of it: "In my opinion [it is] one of the most ingenious experiments that have ever been attempted in the whole domain of physics."

Well, Fizeau's experiment agreed completely with Fresnel's predictions. Michelson later checked his results and found them to be correct.

And thirdly, in 1842, Christian Johann Doppler used the stationary ether theory to postulate that the apparent frequency of light waves (or the colour of the light) emitted by a source or perceived by an observer in motion relative to the ether must differ from the "actual" frequency, when the observer and source are at rest with respect to the ether. A study of stellar spectra soon confirmed this prediction.

Here, broadly, is how the Doppler effect works in the framework of the stationary ether theory.

A rather unsophisticated analysis of the Doppler effect.

1. Receiver and source both at rest relative to the ether. The light from the source enters the receiver with a frequency ω.

2. Source at rest relative to the ether, receiver moving with a velocity v. The apparent frequency reaching the receiver is ω' ($\omega' > \omega$ when the receiver is approaching the source; $-\omega' < \omega$ when it is receding from the source).

3. Receiver at rest, source moving with the velocity v. The apparent frequency of the light entering the receiver is ω'' ($\omega'' > \omega$, but $\omega'' \neq \omega'$, although the relative velocity of the source and the receiver is the same as in the previous case).

This final statement is very important. For if the stationary ether theory is valid, then even if the relative velocity of the light source and the receiver is the same, the apparent frequency as perceived by the receiver is different, depending on whether it is the source or the receiver that is in motion relative to the ether.

Without digressing too far from our subject, we can note that, according to Doppler, the change in the apparent frequency of light waves is exactly analogous to the corresponding effect with sound waves. This is natural, since sound waves are propagated by a prototype stationary ether, namely, the atmosphere.

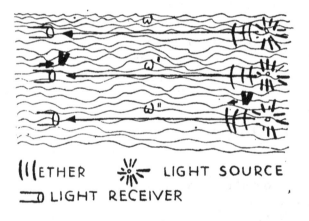

(((ETHER   ☀ LIGHT SOURCE
▭ LIGHT RECEIVER

For a good number of reasons, of which only two will be mentioned, we shall dwell a little longer on the Doppler effect.

Firstly, the Doppler effect plays a very important part in many spheres of physics. In particular, it is one of the most powerful experimental tools of modern astrophysics. Secondly, many people seem to have a very vague idea of the effect, though the principle is very easy to grasp.

Let us solve a problem belonging to the sixth or seventh form of a secondary school. It offers a comprehensive explanation of the Doppler effect for sound and it would have presented an exact analogy of the effect for light waves, if the stationary ether theory were correct.

A marine analogy.

And so, we have a port A. Sailing from it at a speed v is a ship B. The speed of the ship is, naturally, measured with respect to the water. For some reason or other the ship communicates with the port in the following not very convenient way.

At time intervals Δt the port master dispatches messenger boats to the ship. The skipper, too, sends messenger boats to port at time intervals Δt. We denote the velocity of the boats relative to the water by the symbol c. Naturally, $c > v$, otherwise none of the boats from port would ever catch up with the ship.

Let us determine the time interval between two successive arrivals of the boats from the port to the ship and the interval between two successive arrivals of the boats from the ship to port.

First, find the time it takes for a boat to reach the ship from port. If, when the first boat left port, the distance to the ship was a, the time t' it took the boat to reach the ship is found from the obvious equation

$$S = ct_{en\ route} = a + vt_{en\ route}$$

whence

$$t_{en\ route} = \frac{a}{c - v}.$$

When the second boat leaves port the ship will be at a distance $a + \Delta t v$, and it will reach the ship in

$$t'_{en\,route} = \frac{a + \Delta t v}{c - v}.$$

If the first boat was dispatched at time t_0, and the second, at time $t_0 + \Delta t$, the respective moments of their arrival to the ship will be:

$$t_{arrival} = t_0 + \frac{a}{c - v},$$

$$t'_{arrival} = t_0 + \Delta t + \frac{a + \Delta t v}{c - v},$$

and the time interval between the two arrivals will, apparently, be

$$\Delta t_{arrival} = t'_{arrival} - t_{arrival} = \Delta t \left(1 + \frac{v}{c - v} \right).$$

Introducing the notation $\beta = \dfrac{v}{c}$, we have

This equation is worth noting and, furthermore, comparing with the next one.

$$\Delta t_{arrival} = \Delta t \left(1 + \frac{\beta}{1 - \beta} \right) = \frac{\Delta t}{1 - \beta}.$$

As you see, the interval between the two arrivals of the boats is greater than the interval between their respective departures. This is only natural, as the second boat had to travel a greater distance than the first.

Note now that the expression for Δt arrival does not include the quantity a, i.e., the initial distance of the ship from port. In other words, for any two successive messenger boats heading for the ship the increase of the time interval between their arrival is determined only by the relation

$$\frac{v}{c}.$$

If the ship is approaching port all we have to do is change the sign of the speed of the ship. The principle of the solution remains the same (I am sure that the inquisitive reader can establish this for himself).

Thus,

$$\Delta t_{arrival} = \frac{\Delta t}{\left(1 \pm \beta\right)},$$

the plus and minus signs corresponding to the cases of the ship approaching or leaving port, respectively.

If we now introduce a new parameter, namely, the frequency with which the messenger boats are dispatched and arrive, its notation being $v = \frac{1}{\Delta t}$, we obtain

$$v_{arrival} = v_{dispatch}\left(1 \pm \beta\right).$$

This example demonstrates how the apparent frequency of sound waves changes when their source is at rest relative to the atmosphere and the receiver is moving. If the stationary ether theory were correct, the case should be exactly the same with electromagnetic waves. The reader, I am sure, will be able to calculate for himself the frequency with which the messenger boats dispatched from the ship arrive in port and will obtain the equation

$$v_{arrival} = \frac{v_{dispatch}}{\left(1 \pm \beta\right)},$$

where the plus sign corresponds to the ship leaving port and the minus sign, to the ship approaching port.

As you see, although qualitatively the frequency changes in the same way in both cases, quantitatively the results differ depending on whether the source or the receiver is moving with respect to the ether, even if their respective velocities relative to the ether are the same.[†]

It is often declared that the Doppler effect can be observed directly by listening to the whistle of an approaching train. I must say that this may be possible only for people with a very keen ear. Usually, though, what is observed is not a change in frequency (pitch) but a change in intensity (volume). Actually observers with no special gift for music, who are rather "spoilt" by education, regard the change in intensity as the theoretically predicted change in frequency and come to the conclusion that acoustically the Doppler effect takes the form.

[†]In the *first* example the port is the source of light, the ship is the receiver, the messenger boats are the light waves, and the sea is the stationary ether.

In the *second* example the ship is the source and the port the receiver.

Actually, though, it is volume not pitch that is laid off on the ordinate axis. The curve characterising the change in frequency, which is usually imperceptible to the ear, is given by the following drawing:

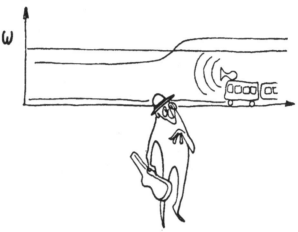

Here ω is the "actual" pitch of the whistle (that is, the pitch of the whistle when both source and observer are at rest relative to the atmosphere).

When the speed of the train reaches about 65 km/hr, the change in pitch is about half a tone (instead of C we should hear C sharp). But as a train whistle rarely produces a "pure" (monochromatic) sound, the observed picture is somewhat more complex. I can only repeat that without special laboratory instruments hardly any observer not possessing an excellent musical ear can hope to register the Doppler effect. To conclude, the actual picture is described by more complex equations than those given here.[‡]

At least one advantage of being music-minded.

[‡]Well, the TGV trains in France now travel with the speed ~ 300 km/hr, and then the change in pitch would be clearly detectable for anybody. But the TGV trains do not whistle... — A.S.

We have considered cases in which the velocity is directed along a straight line joining the source and the receiver. When this is not so (which is usually the case) instead of the total velocity its projection on the line joining the source and the receiver is taken. And one final remark. As is observed in the last drawing, at the moment when the train passes by the observer, and the projection of the velocity on the straight line drawn from the train to the observer is zero, the apparent frequency equals the actual frequency.

Now we can consider some interesting consequences which arrise when the Doppler effect is applied to light waves.

When the receiver and the source are approaching each other, the apparent frequency increases. If we travelled towards a star with a velocity comparable with that of light, we would see the visible portion of its radiation spectrum so displaced as to make its infra-red, and even radio-wave, radiation visible to the eye. Therefore, theoretically at least, it would be quite possible to observe a spectacular glow about the mast of a radio station if we approached it with a velocity close to that of light.

On the other hand, if we receded with sufficient speed from a source of radiation, we could be able to see gamma-quanta with our own eyes. In such a case an atomic pile would be a brilliant light source.

I once read a sci-fic book about a surgeon who found a way of changing the retina of the eye so that it could see very long electromagnetic waves. The Doppler effect makes such things possible without surgical interference.

What has just been said holds good for the correct theory of the Doppler effect based on relativistic conceptions. In fact, the Doppler effect is "almost correct" in the stationary ether framework. There are some substantial differences, however.

We have seen that within the framework of the stationary ether the following cases are possible: (1) receiver approaches source with velocity v relative to the ether; (2) source approaches receiver with velocity v relative to the ether. In both cases the frequency increases, but to a different degree. Referring to the equations given before, you will readily notice that the difference between the apparent frequencies is of the order of magnitude β^2.

We shall see later on that in the theory of relativity it is meaningless to speak of the world ether — or anything else — as an absolute frame of reference. The frequency change is completely determined by the relative velocity of the source and receiver, increasing when they are approaching and decreasing when they are receding. The formula for the frequency change, however, is different, namely:

$$v' = v\sqrt{\frac{1 \pm \beta}{1 \mp \beta}}.$$

The correct formula for the change in frequency of light waves. We shall find it again in Chapter XIV.

Secondly, precise theory of the Doppler effect based on Einstein's theory leads to the conclusion that the apparent frequency must change even if the projection of the velocity on the straight line joining the source and the receiver is zero (in our case, when the train is passing by the observer). This interesting conclusion, the so-called transverse Doppler effect, is linked closely with the change of time-flow in different reference frames. Einstein himself regarded experimental confirmation of this theoretical prediction as an important argument in favour of the theory.

Unfortunately, the author is at a loss to set forth in simple form the essence of the Doppler effect from the point of view of relativity. In future, therefore, we shall restrict ourselves to brief remarks concerning the Doppler effect. The basic features of the phenomenon have been outlined in the foregoing.

To return now to the ether, it could be claimed that, qualitatively, the theoretical explanation of the Doppler effect based on the stationary ether concept was confirmed by experiment. Nineteenth-century physics was unable to offer a precise analysis since the actual frequency formula differs from the one based on the stationary ether concept by a quantity of the order $\left(\dfrac{v}{c}\right)^2$.

Thus, despite all the snags encountered by the ether theory, many facts seemed to testify to its validity.

The aberration of light, experimental confirmation of Fresnel's partial-drag hypothesis, the Doppler effect and, finally, practically the whole of the wave theory all served as seemingly sound arguments in favour of the stationary ether.

It is often claimed that a crucial test of the validity of a theory lies in a correct prediction of the results of new experiments. The example of Fresnel's theory demonstrates that the utmost caution is required in such questions. Even if a theory is confirmed by experiments, there is no absolute guarantee of its validity until a sufficient experimental data are accumulated. After all, Fresnel did predict the outcome of Fizeau's experiments!

On the other hand, though, a single experiment is sufficient to overthrow a theory. And in 1881, Michelson finally carried out the first experiment which made it possible to detect the effect of the Earth's motion relative to the ether "in the second order of the ratio $\dfrac{v}{c}$". The result was negative.

The Earth's motion relative to the ether did not affect optical phenomena in "the second order of the ratio $\dfrac{v}{c}$"!

And finally, Michelson's experiment, which spelt the doom of the ether.

Something of an in-
structive lyrical digres-
sion.
At this juncture tradition prescribes the author and reader to observe a moment of solemn silence.

Now we enter a dank basement at the Potsdam astrophysical laboratory. The lights outside have long been extinguished, the burghers are sound asleep. Albert Abraham Michelson has just completed the adjustment of his apparatus. He tarries a moment, then with trembling fingers switches on the light source and turns hastily to the eyepiece of the interferometer.

Hardly more than a few seconds pass, but to him it seems that an eternity has lapsed while his eyes strain to see the expected shift of the interference bands. And another eternity passes before he realises that the effect is zero.

His first feeling is not of joy but of utter exhaustion. Joy will come later, he knows that as sure as he knows that he has just completed an experiment without parallel in the history of physics.

The author confesses that he once used to picture research work and the making of a major scientific discovery in such a vein. Our imagination is easily captivated by spectacular scenes, which usually linger longest in our memory. Newton's apple, Janos Bolyai's duels, Galileo's recantation, the death of young Evariste Galois, and Archimedes' "Eurika" are the sundry (and, alas!, often only) pictures that we conjure up in our mind when we speak of scientists and their work. In short, the true romance of science, the romance of daily drudgery, is hardly recognised despite vain appeals to the contrary. This, to be sure, refers not only to science. The conquest of Everest appeals to our imagination much more

than an ascent of some unknown peak, even
if it required as much grit and stamina as to
conquer the world's tallest mountain. The very
fact that people can seriously engage in paltry
discussions as to who in fact was the first to
set foot on the summit, Tenzing or Hillary,
is hardly to the credit of many a mountain-
climbing "enthusiast". Similarly, many "science
enthusiasts" are enthused not by the work but
by spectacular "sound effects".

Many people claiming "an interest in
physics" know of Michelson not as of one of the
most hard-working and subtle experimenters
in the history of science but only as the maker
of the experiment which led to the development
of the relativity theory. The result, and only

An interesting comment by Michelson on his experiments.

the result, is what has surrounded Michelson with a halo of sanctity in the minds of many a "cultured" person.

But here is what Michelson himself wrote of his experiment many years later:

"It was considered that, if this experiment gave a positive result, it would determine the velocity, not merely of the Earth in its orbit, but of the Earth through the ether. With good reason it is supposed that the Sun and all the planets as well are moving through space at a rate of perhaps twenty miles per second in a certain particular direction. The velocity is not very well determined, and it was hoped that with this experiment we could measure this velocity of the whole solar system through space. Since the result of the experiment was negative, this problem is still demanding a solution.

"The experiment is to me historically interesting, because it was for the solution of this problem that the interferometer was devised. I think it will be admitted that the problem, by leading to the invention of the interferometer, more than compensated for the fact that this particular experiment gave a negative result."

This appraisal was prompted not only by the modesty of a great man. Michelson was very disappointed at the negative result of his experiment. He had hoped simultaneously to establish the motion of the solar system with respect to the fixed star frame and confirm the stationary ether theory.

He achieved neither the one nor the other. All the experiment did was to fail to confirm the hypothesis of the stationary ether. It was still a long way to the formulation of relativity theory, and Michelson could only state that the result of his experiment was totally incomprehensible. His disappointment, therefore, was sincere and natural. His consolation was that he had invented a truly remarkable instrument, the interferometer.

It is worth having a closer look at the idea of Michelson's experiment as well as at its theoretical basis. The exact theory of the experiment, incidentally, bears a rather distant resemblance to conventional descriptions of the set-up. Suffice it to say that in his first announcement Michelson presented an erroneous calculation.

A correct computation of the anticipated effect on the basis of the stationary ether hypothesis gives half of Michelson's estimate.

In Michelson's words the idea of the experiment was Maxwell's, while the scheme (mind you, the scheme) of the apparatus was extremely simple.

According to the stationary ether theory, the velocity of light relative to the ether does not depend on the motion of the light source with respect to the ether (just as the velocity of sound in the atmosphere does not depend on the motion of a sound source relative to the atmosphere).§ Therefore, if the stationary ether hypothesis is correct, the following interesting effects should be observed.

Consider a source of light and a mirror at rest with respect to each other. Like everything else in the universe, they are submerged in the

---

§As pointed out before, the stationary ether theory is similar to the theory of the propagation of sound waves in the atmosphere, which can be regarded as a "stationary acoustic ether".

ever-calm ether sea. It will be readily observed that the time it takes a beam of light to travel from the source to the mirror and back will differ depending upon whether the frame is at rest or in motion with respect to the ether.

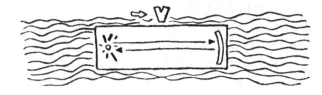

Michelson's experiment was based exactly upon this consideration. Within the framework of the stationary ether theory, it can be described as follows.[1]

A description of Michelson's experiment "translated" into a school problem about swimmers.

A square raft is towed across still water (we assume the raft to be square in order to simplify subsequent computations). Its speed relative to the water is v. From corner A of the raft two men, swimmer No. 1 and swimmer No. 2, dive into the water. Both swim with the same speed c.

Swimmer No. 1 swims towards corner D, swimmer No. 2, towards corner B. On reaching their respective goals they turn back to A. Obviously, $c > v$, as otherwise the swimmers would be unable to keep up with the raft.

A bit of simple mathematics.

Our task is to determine the time it takes either man to swim back and forth. The problem, you see, belongs to secondary-school algebra and, with your kind permission, I present the solution without any further explanations.

[1]Since the stationary ether theory, as we shall see later, is wrong, the interpretation of the experiment set forth here is also all wrong.

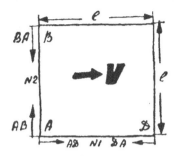

For swimmer No. 1:

(i) $t_{ADA} = t_{AD} + t_{DA}$;

(ii) $ct_{AD} = l + vt_{AD}$, $t_{AD} = \dfrac{l}{c-v}$;

(iii) $ct_{AD} = l - vt_{AD}$, $t_{AD} = \dfrac{l}{c+v}$;

(iv) $t_{ADA} = \dfrac{l}{c-v} + \dfrac{l}{c+v} = \dfrac{2lc}{c^2-v^2} = \dfrac{2l}{c} \times \dfrac{1}{\left(1 - \dfrac{v^2}{c^2}\right)}$.

Here $\dfrac{2l}{c} = t_0$ is the time it would have taken the swimmer to swim there and back if the raft were at rest.

If $\dfrac{v}{c} \ll 1$, then $\dfrac{1}{1 - \dfrac{v^2}{c^2}} \approx \left(1 + \dfrac{v^2}{c^2}\right)$.**

**We have employed here for the first time an approximate solution. Though of great importance, hardly any attention is given to approximate solutions in school. Therefore the development of the last equation, as well as of another one which we shall use several times later on, should be explained.

If a quantity α is very small, we may accept that

$$\frac{1}{\sqrt{1-\alpha}} \approx 1 + \frac{\alpha}{2}.$$

Then the time it took swimmer No. 1 to swim back and forth is

$$t_1 = t_0 \left(1 + \frac{v^2}{c^2} \right).$$

This is easily proved.

(1) When α is small, $\sqrt{1-\alpha} \approx 1 - \frac{\alpha}{2}$. Indeed, squaring both sides of the approximate equality, we obtain

$$1 - \alpha \approx 1 - \alpha + \frac{\alpha^2}{4}.$$

The right-hand term is greater than the left side by $\frac{\alpha^2}{4}$. But if $\alpha \ll 1$, then α^2 is quite small and can be neglected (for example, if $\alpha = 0.001$, then $\alpha^2 = 0.000001$). Hence, to an accuracy of the order α^2

$$\frac{1}{\sqrt{1-\alpha}} = \frac{1}{1 - \frac{\alpha}{2}}.$$

(2) Multiply the numerator and denominator of the fraction $\dfrac{1}{1 - \dfrac{\alpha}{2}}$ by $\left(1 + \dfrac{\alpha}{2} \right)$. We obtain

$$\frac{1}{1 - \frac{\alpha}{2}} = \frac{1 + \frac{\alpha}{2}}{1 - \frac{\alpha^2}{4}}.$$

As before, we can neglect the term $\dfrac{\alpha^2}{4}$ in the denominator, and finally

$$\frac{1}{\sqrt{1-\alpha}} \approx \frac{1}{1 - \frac{\alpha}{2}} \approx 1 + \frac{\alpha}{2}.$$

The accuracy of this equality is of the order of α^2. The fact that we have neglected terms of the order α^2 twice need not cause apprehension as it hardly increases the error.

For swimmer No. 2 the solution is a bit more involved. The shortest path from A to B is the hypothenuse of a right triangle ABB', where B' is the position of corner B by the time the swimmer reaches it. If swimmer No. 2 is clever enough he will take into account the motion of the raft as soon as he dives into the water and will swim along the hypothenuse. The same goes for his path from B back to A. The time is found simply:

(i) $t_{ABA} = t_{AB} + t_{BA} = 2t_{AB};$

(ii) $c^2 t_{AB}^2 = l^2 + v^2 t_{AB}^2, \quad t_{AB}^2 = \dfrac{l^2}{c^2 - v^2};$

(iii) $t_2 = t_{ABA} = t_0 \dfrac{1}{\sqrt{1 - \dfrac{v^2}{c^2}}}.$

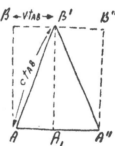

Again, if $\dfrac{v}{c} \ll 1$, then

$$\frac{1}{\sqrt{1 - \dfrac{v^2}{c^2}}} \approx \frac{1}{1 - \dfrac{v^2}{2c^2}} \approx \left(1 + \frac{v^2}{2c^2}\right).$$

And finally we have

$$t_2 = t_0 \left(1 + \frac{v^2}{2c^2}\right).$$

Note that this time is less than that of swimmer No. 1.

You see that

$$t_1 - t_2 = \frac{t_0 v^2}{2c^2}.$$

Swimmer No. 1 is at a disadvantage and returns to A later than swimmer No. 2. If the raft turns through 90° without changing the direction of motion, the swimmers will change parts and No. 2 will come in later than No. 1.

And now all we have to do is to substitute (i) the stationary ether for the water, (ii) Michelson's apparatus, travelling through the ether together with the Earth, for the raft, (iii) beams of light for the swimmers, and we have the scheme of Michelson's experiment.

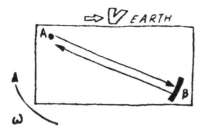

Conclusions. The theoretical aspect of the experiment has been described.

The analogy is exact. In our example we have strictly outlined the elementary theory of Michelson's experiment from the point of view of the stationary ether hypothesis. Actually, though, the case is more involved due to aberration and refraction of light in optical instruments.

Thus, in order to demonstrate the Earth's motion through the ether, we must take a source of light and a mirror and measure the time it takes a beam of light to travel back and forth (see drawing). When the platform of the apparatus is turned we should, as demonstrated, observe a change in the time it takes the light beam to travel the same distance.

The time is greater when the path AB is parallel to the Earth's motion through the ether and smaller when it is perpendicular to the line of motion (when the "ether wind" only "blows" the light beam slightly off course). If we can observe this difference then we have proof of the Earth's motion through the ether. All is very simple.

True, we may well doubt whether Michelson's experiment was really so "simple", for the anticipated time difference was 1/100,000,000th of the time required for the light beam to travel through the apparatus (several yards),[††] which also was no greater than a hundred millionth of a second.

Remarks on the practical implementation of the experiment.

Maxwell considered the practical implementation of his idea to be a hopeless task, and not without reason. For the relative accuracy of measurement (10^{-8}) was equivalent to about one second in several thousand years. Or take another comparison: the time difference which Michelson undertook to detect was less than the time it takes an electron to circle an atomic nucleus. The formidability of the obstacles confronting Michelson are difficult to imagine. The base of his instrument was about one metre long. In order to detect a time difference with an accuracy of 10^{-8}, one had to be quite sure that the distance travelled by the beam was constant at least with an accuracy of 10^{-9}, otherwise the time it took the beam to travel that distance could change simply because of the change in distance. And an accuracy of 10^{-9}

[††] As $v_{earth} = 30$ km/sec,

$$\frac{v_{earth}}{c} = 10^{-4}, \text{ and } \left(t_1 - t_2\right) = \frac{t_0 v^2}{2c^2} = \frac{t_0}{2} \times 10^{-8}.$$

is of the order of about 10 angstrom units in one metre! In case you don't know, 10 angstrom units is about the length of three or four atoms in a row.

Obviously, the slightest jolt or temperature change would cause a much greater change in the base. Literally, Michelson could not even breathe on his apparatus! To exclude external vibrations, he mounted it on a stone slab which rested on a round wooden plaque floating in mercury. The mercury was contained in a bowl standing on a pillar dug into the floor of a basement room. Vibrations were thus excluded, but how to measure the time it took the light beam to travel back and forth? Any attempt to do so directly would, of course, have doomed the experiment to failure. Michelson very ingeniously made use of the interference effect.

If a light beam is split into two equal parts and the two beams are later reunited they will project on a screen a peculiar pattern of interference lines. The drawing shows the way in which Michelson split the light beam. A glass

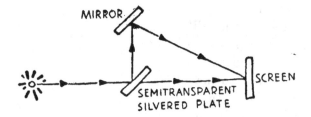

plate covered with a thin semi-transparent layer of silver reflects part of the incident light and lets through the rest. The waves of the two beams form an interference pattern on the screen. The difference between the paths travelled by them being constant, the interference pattern should not change, insofar as it depends only on the relative time lag between the two beams. If this relative time lag is changed slightly, the interference pattern also changes. The value of this "slight change" is in the incredible accuracy range of the order 10^{-10}.

Michelson decided to make use of the above considerations in his experiment. In his apparatus he split a ray of light into two mutually perpendicular beams and then made them reunite. In the eyepiece of his interferometer could be observed a certain interference pattern. As long as external conditions remained unchanged, so did the interference pattern. As a matter of fact, Michelson observed it unchanged for several hours running.

Now, if the stationary ether theory is correct then, as we have seen, light beams propagating through the ether parallel to the Earth's path and perpendicular to it are placed in different conditions and the time required for them to travel the same distance is not the same. Accordingly, a turn of the apparatus through a right angle (swimmer No. 1 and swimmer No. 2 change parts) should have caused a change in the interference pattern. But...

A more or less faithful description of Michelson's experiment.

In his very first experiment Michelson failed to observe any regular change in the interference pattern when he turned his instrument through 90°. This was contrary to all theoretical expectations.

Since the problem concerned was such a fundamental one as the ether theory, which, it seemed, had been verified so many times, the negative result of the experiment immediately raised doubts as to the accuracy with which it had been conducted.

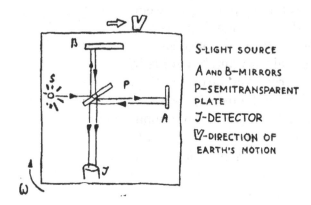

S-LIGHT SOURCE

A AND B—MIRRORS

P—SEMITRANSPARENT PLATE

J-DETECTOR

V-DIRECTION OF EARTH'S MOTION

Incidentally, Academician Vavilov points out that the accuracy of Michelson's first experiment was actually insufficient and that Michelson rather guessed intuitively than strictly substantiated its inferences. That is why Michelson set about to verify his observations.

Six years later, in co-operation with Morley, he repeated the experiment, using a much better apparatus. This time there seemed to be no doubt as to the absence of the expected effect. Yet doubts continued to be voiced.

A comment on the character of physicists.

In general, physicists are usually very cautious in their appraisal of findings of such importance as Michelson's. So the experiment was repeated again and again with improving accuracy right up till 1927!

The final verdict based on the sum total of the tests was that Michelson was right: the Earth's motion through the ether produced no apparent effect and there was no "ether wind". Note that this was 1927, forty years after Michelson's first experiment and 22 years after the relativity theory was enunciated. Scores of tests had already been carried out which confirmed the relativity theory, and yet scientists continued to verify Michelson's findings.

Such scrupulous exactingness is quite in keeping with the character of physicists. Every general proposition in the history of physics has been subjected to the severest acid tests. Often it is impossible to name the blessed date when a theory is finally declared to be definitely sound.

The experiment suggested that there was something wrong with the stationary ether hypothesis. This was Michelson's conclusion, but he could not say what was wrong. Maybe the "ether wind" blew at the Earth's surface? After all, the experiments had been carried out in a basement. Michelson allowed for such a possibility. It was useless, he wrote, to attempt to solve the question of the motion of the solar system by observing optical phenomena at the surface of the Earth; the possibility could not be excluded that even at some moderate altitude above sea level, say on some lone mountain top, the relative motion could be detected with the help of an apparatus along the lines described in his experiments.

More doubts.

Michelson's experiment was later repeated on a mountain top, and even in a balloon, and the result remained negative. Several times doubts were raised concerning the accuracy of the computations and data processing, and Michelson's work was reviewed again and again until physicists were convinced of the absence of any "ether wind".

Many other experiments "of the second order" based on different principles were carried out, but they too produced a negative result. Relativity theory had been postulated and everything set right, the ether had been thrown "into the same dustbin as phlogiston, thermogen and *horror vacui*,"‡‡ as one scientist aptly formulated the case at the turn of the century, but experimenters continued to verify Michelson's findings. It is difficult to pinpoint the exact date when this verification ceased to be of scientific interest.

It is inevitable that a moment arrives when critical suspicion of a new theory, quite natural at first, turns into ossified conservatism. Relativity theory, in any case, has been subjected to such "third degree" scrutiny and appeared before the competent judge of experiment so many times that today there can be no doubt as to its absolute integrity.

Summing up the situation.

Let us tarry for a moment and see what we have achieved. We have traced, very superficially, the development of the ether theory and found, as a result of Michelson's experiment (or rather his second work jointly with Morley in 1887), that it required some drastic revision. We are still in the dark as to the nature of this revision. For, although the ether hypothesis has reached an impasse, it still provides a very satisfactory explanation of many facts. If the ether hypothesis appeals to you in any way you are all the more in a position to realise why the downfall of the ether meant a revolution in physics.

‡‡Phlogiston, thermogen and *horror vacui* (fear of vacuum) used to be highly popular theoretical concepts which were later discarded as insolvent.

From our point of view the hypothesis of the ether — a very strange substance — is of purely historical interest. A realisation of the value of the ether to physicists will enable us to understand better Einstein's achievement.

The theory of relativity can be examined without even mentioning the ether. In fact, it would be even easier to understand Einstein's postulates. It would be a pity, though, to lose sight of the perspective. I mentioned at the beginning of the book that Einstein's postulates are simple. Allow me now to retract my words. Einstein's theory is harmonious and graceful. His postulates are probably much more natural and formulated much more precisely and clearly than the whole of classical physics. His theory provides an admirable explanation to all known phenomena and experimental findings.

A few more words about Einstein himself.

Finally, relativity theory makes immediate use only of experimental data and in this sense is based on direct experiment.

With all that, however, I personally am at a loss to explain how twenty-six-year-old Albert Einstein ever arrived at his theory. The consideration that, after Michelson's work, relativity theory was the only alternative is unconvincing.

There were many ways to patch up the ether theory. Moreover, they were used and some successes were scored. Lorentz, for example, sought to explain Michelson's experiment with the help of the ether and practically all the fundamentals of classical physics. Ritz developed a theory without the ether but with the whole of classical mechanics intact. From the point of view of his time Einstein took to the most unlikely path.

Undoubtedly, the creation of the relativity theory is due primarily to the inimitable qualities of its maker, which can be named but which defy explanation or comprehension.

I think that attempts to analyse in detail the working of genius occupy a prominent place among many useless pursuits. As for Einstein himself, he says that he had puzzled over these questions for some ten years. His exact words will be quoted in the next chapter, but I should like here to draw attention to the wonderful simplicity with which Einstein writes: "I was pretty much convinced of the validity of this before...."

And to end with the ether, here is Michelson's conclusion, which reflects fairly accurately the state of affairs on the eve of the relativity theory:

"A number of independent courses of reasoning lead us to the conclusion that the medium which propagates light waves is not an ordinary form of matter. Little as we know about it, we may say that our ignorance of ordinary matter is still greater.

Here Michelson probably cites Lord Kelvin, the most quick-witted physicist in the history of science.

"...The phenomenon of the aberration of the fixed stars can be accounted for on the hypothesis that the ether does not partake of the Earth's motion in its motion about the Sun. All experiments for testing this hypothesis have, however, given negative results, so that the theory may still be said to be in an unsatisfactory condition."

CHAPTER XI,

in which the author seeks to confuse the patient reader by convincing him of the contradictions of Einstein's postulates. As a result, it turns out that they are incompatible with classical mechanics, and the author asks the reader to share his profound admiration for Einstein. The first half of the chapter may seem somewhat difficult, but the reader may find consolation in the fact that it is the second half that matters more

EINSTEIN (basic postulates)

Newton was the most fortunate among scientists, because the system of our world can be created only once.

LAGRANGE

At long last we are approaching our goal. All that follows is concerned directly with Einstein's theory. We shall not go deep into other attempts to explain Michelson's result, although they are both interesting and edifying. Still a few words should be said about Einstein's predecessors, if only to see once again how many pathways present themselves as soon as an old theory runs into difficulties and a new one has to be produced.

The first is Lorentz, who worked a lot on the electromagnetic field theory and who developed, in the eighteen-eighties, the most rational and

The usual general remarks. A few words about Einstein's predecessors.

progressive system of "ether physics". After Michelson's findings he made a last-ditch attempt to save his theory in 1904.

Lorentz assumed that all bodies moving relative to the ether contract in the direction of the motion according to the formula

$$l = l_0 \sqrt{1 - \frac{v^2}{c^2}},$$

where l_0 is the length of the body at rest relative to the ether, and v is its velocity relative to the ether.[*]

Lorentz even produced a very plausible (though purely hypothetical) explanation of the phenomenon on the basis of his theory of the structure of matter. Lorentz's theory not only explained the results of the Michelson experiment: its formal, mathematical structure bore a close analogy to Einstein's theory.

Even closer to the relativity theory approached the ideas of the eminent French mathematician Poincaré.[†]

People often wonder why Lorentz, and especially Poincaré, who came so close to the relativity theory, failed to make the final step. Tradition requires an explanation of this.

The theory of relativity was developed by Einstein, and not Poincaré or Lorentz, for the sole reason that Einstein proved more capable of getting down to the essence of the matter. This reply resolves the problem completely. To tell the truth, however, the widespread notion that Poincaré and Lorentz were practically at

---

[*]A similar hypothesis was formulated independently by Fitzgerald.

[†]An interesting fact: Poincaré's paper went to press only three weeks after Einstein's.

the verge of formulating the relativity theory is simply wrong.

Any physical theory is first and foremost determined not by its mathematical props but by its physical content. It is true that Lorentz, and especially Poincaré, came close to a mathematical formulation of the theory, but they failed to grasp its physical aspects. It was this last step that was most difficult, and it is hopeless to speculate how long it would have taken Poincaré to arrive at Einstein's ideas.

A small contribution to history.

Einstein's paper *On the Electrodynamics of Moving Bodies* appeared in the seventeenth volume of *Annalen der Physik*, a sedate German scientific journal, in 1905. We need not go into the content of this work. An excellent description of it is given by Leopold Infeld:

"The title sounds modest, yet as we read it we notice almost immediately that it is different from other papers. There are no references; no authorities are quoted, and the few footnotes are of an explanatory character. The style is simple, and a great part of this article can be followed without advanced technical knowledge. But its full understanding requires a maturity of mind and taste that is more rare and precious than pedantic knowledge. Even today its presentation and style have lost nothing of their freshness. It is still the best source from which to learn relativity theory. Its author was an outsider, not even a member of the scientific profession. He was, in 1905, a young Ph.D., twenty-six, and a clerk in the Patent Office in Berne, Switzerland."

Einstein began with sorting out the indubitable experimental facts. Facts are without number, but as often as not they contradict one another. To sift away the irrelevant and retain the fundamental is in itself a formidable task.

Here is something indubitable: Michelson's experiment proved that optical phenomena on Earth do not depend on its motion relative to the fixed stars. The Earth's motion with respect to the stars can, with sufficient accuracy, be regarded as uniform and rectilinear (which is very important), which means that the *uniform and rectilinear* motion of the Earth relative to the fixed stars has no effect on optical phenomena on Earth.[‡] But if that is so, then Galileo's relativity principle is valid for electromagnetic phenomena and it may be a general law of nature! Einstein takes this proposition as the first postulate of his theory.

_____

[‡]Attention should be drawn to the latter statement, as the rotational motion of a reference frame relative to the fixed stars does affect optical and electromagnetic phenomena within that frame. In this connection, the following hypothetical oddity is of interest. It follows from relativity theory that the diurnal rotation of the Earth on its axis should influence optical phenomena. In 1925, Michelson and Gale staged an extremely subtle experiment which revealed this effect. As far back as 1913, the effect of the rotation of a reference frame relative to the fixed stars on electromagnetic phenomena was demonstrated experimentally by Sagnac (the idea of the experiment was also Michelson's).

But the joke is that, for these experiments, the predictions of relativity theory were essentially the same as the predictions of the stationary ether theory. In this case the predictions of a correct and a false hypothesis were the same.

You can imagine how much later Einstein's theory would have appeared if Michelson had first carried out his experiment with the diurnal rotation of the Earth in order to demonstrate that the ether was not carried along by the Earth's rotation! For it would have proved convincingly that the stationary ether did in fact exist!

All laws of nature are the same in all inertial reference frames moving uniformly in a straight line relative to each other.

You observe that the only difference between this postulate and Galileo's principle is, essentially, that it speaks of "the laws of nature" instead of "the laws of mechanics". Accordingly, the physical content is also essentially the same as that of Galileo's relativity principle, with the important amendment that now the equality of inertial frames with respect to all physical laws, and not only the laws of mechanics, is postulated.

We are already familiar with the physical meaning of the relativity principle. Namely, uniform motion of a reference frame in a straight line relative to the fixed stars has absolutely no effect on anything. (No experiment carried out inside a closed room can ever show that it is in uniform rectilinear motion with respect to the fixed stars.)

The relativity principle may possibly seem trite in this form. After all, the stars are so far away that intuitively it seems clear that they can have no effect on anything. But recall the case of rotation. If a totally isolated room is caused to rotate, an observer inside can detect this motion immediately. Hence, the relativity principle is not so self-evident. More, it is a very

The first postulate in Einstein's theory: the relativity principle. At this point it would be worthwhile reviewing Chapter V.

And what if Michelson would have been satisfied with this finding? He might never even have undertaken his "real" experiment! Not a bad topic for debating what would have happened if....

But, if anything, all the foregoing offers graphic illustration to the significant fact that a single experiment may be sufficient to overthrow a whole theory. To assert a new theory no end of tests and verifications are required.

remarkable principle, but then, that is how the world is made.

Thus, Einstein extends the relativity principle to all laws of nature (and first and foremost to the laws of electromagnetism), thereby providing an explanation for the zero effect in Michelson's experiment: it is an obvious result of the relativity principle.

Uniform motion in a straight line relative to the fixed stars affects nothing. Therefore, even though the Earth is moving, the light beams in Michelson's apparatus behave exactly as though it were at rest. Without going into the details, the following is a close analogy.

When two men play billiards in the saloon of a uniformly moving steamship everything takes place exactly as if they were on solid ground. It makes no difference to the billiard

The relativity principle and Michelson's experiment.

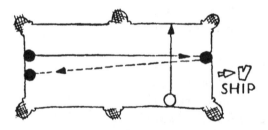

balls whether they are propelled in the direction of the motion or at right angles to it. A ball shot across the billiard table in the direction of the ship's motion will take as long to reach the opposite side as to get back (neglecting, of course, the retardation of its speed due to friction on the cloth). Neither does a ball shot perpendicular to the ship's motion "realise" that the ship is moving: it hits the opposite side precisely where it was aimed and the ship's motion does not deflect it. In short, the billiard players are in no way discomfited by the fact that they are playing on board a ship and not on dry land.

Now imagine Michelson's set-up instead of the billiard table, with light beams instead of the billiard balls and the Earth instead of the ship: obviously the experiment must take place exactly as if the Earth were at rest relative to the fixed stars (or the stationary ether). Like the billiard balls, it does not matter at what angle the light is directed relative to the Earth's motion, and the time it takes to cover a given distance does not depend on that angle.

Evidently, if we accept Einstein's relativity principle, we should give up the idea of a "distinguished" reference frame, such as the stationary ether. If you remember, in Chapter V the question of the existence of an "absolute frame" remained unsolved. We assumed that it might be possible to find such a frame if we studied, say, electromagnetic phenomena. By extending the relativity principle to all laws of nature, Einstein negates the very idea of the existence of some distinguished frame of reference.

The relativity principle and the ether.

But doesn't aberration contradict the relativity principle? So far we explained it on the basis of an absolute reference frame, the stationary ether. But who said that that was the only possible explanation? As such the phenomenon of aberration does not contradict the relativity principle. It is our explanation of aberration that clashes with relativity. Well, all the worse for the explanation. "As such", the results of any experiment show only what they indicate.

Aberration and first doubts.

And so, at the bottom lies the relativity principle. On this score Einstein toes the "classical" line. More, he expands Galileo's classical formula and its region of application. The trouble, though, is that, taken by itself, the relativity principle does little to clarify the situation. The fact that we must give up the ether theory is no cause for concern. We can simply forget about the ether and study all experimental data with an open mind. But here is where the difficulties come in.

A "learned" example.

Let us use the relativity principle to analyse a simple experiment. Consider our old inertial reference frames O and O', whose velocity relative to one another is v. In frame O we stage an experiment to determine the velocity of light. Let us call it "Test L". For this we take a light source S fixed in the O frame and measure the speed of light in some way (say by Fizeau's method). The set-up for our experiment is also fixed in the O frame. We find that the speed of light as determined by us is c.

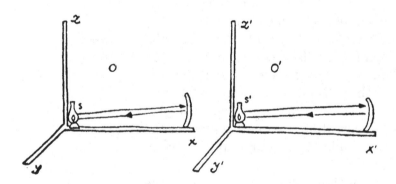

Now let us repeat our set-up in the frame O', which is moving relative to the O frame (see drawing), taking in it a fixed light source S', etc. Labelling our experiment "Test L'", we stage it so that all the conditions relative to the primed frame are identically the same as the conditions of Test L in the unprimed frame.

By the relativity principle, the velocity of light as given by Test L' is also c, since one inertial frame is as good as any other, otherwise we would find that the laws of nature are different in different inertial frames. So far so good.

Now, we are fully entitled to investigate any experiment from any other reference frame. So let us describe Test L' in terms of frame O. In the unprimed frame the light source S' and the whole experimental set-up is moving to the right with a velocity v. Evidently the velocity of the light beam travelling from S' to the right is equal to the velocity of light plus the velocity of the primed frame relative to the unprimed one, i.e., $(c + v)$. Correspondingly, the velocity of the reflected light beam equals the difference between the two, i.e., $(c - v)$.

We have approached a very important stage, so in order to gain a better understanding of what is to come, let us proceed from abstract reasoning to concrete examples.

Let a physicist be travelling with all his instruments in the middle of a carriage in a uniformly moving train. His measurements tell him that the velocity of light relative to the light source does not depend on the direction and is equal to c. In other words, he has established that a light beam reaches the end walls of the

Attention! A major, and very important mystification begins!

carriage simultaneously after a time interval
Δt and has determined the velocity:

$$c = \frac{\frac{1}{2}l}{\Delta t}.$$

Now, if the carriage were made of glass an
observer at the roadside could also investigate
the propagation of light in it. His observations,
however, would be somewhat different. By the
time the light beam travels from the source to
the end walls of the carriage, the train moves
forward some distance. The forward wall tries
to escape from, and the rear wall to approach,
the light beam. Therefore, the beam has to
travel a smaller distance to reach the approaching wall. Yet it reaches the two walls at the
same time! Obviously, this is possible only if
the light travels in the direction of the motion
faster than in the reverse direction.

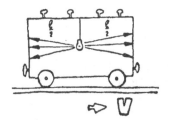 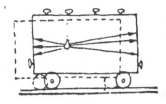

The respective speeds are easily found
and, as just stated, the speed of the "forward"
moving beam is $c + v$, and of the "backward"
moving one, $c - v$, where v is, of course, the
speed of the train.

The same conclusion can be reached by a
different course of reasoning. The speed of
light relative to its source is constant and is

equal to c in any reference frame (the relativity principle!).

If the beam from the locomotive's headlight is "escaping" from the train at velocity c, and the train is receding from an observer at a velocity v, then the beam is moving relative to the observer at velocity $c + v$. Similarly, a light flashed from the tail carriage of the train will have a speed $c - v$ relative to the embankment, which we see immediately when we apply the rule for the composition of velocities.

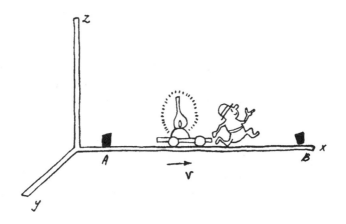

In other words, we have come to the conclusion that the speed of light depends on the motion of the source. To reach this conclusion we made use only of the relativity principle. Hence, if our proposition is not confirmed by experiment, then the relativity principle is not valid as far as electromagnetic phenomena are concerned.

Thus, if we accept the relativity principle we must concede that if a light source is moving at a velocity v in a reference frame, then an

The foregoing reasoning and conclusions are all wrong. Taken by itself, the relativity principle does not lead to the conclusion that the velocity of light depends on the motion of its source. But where is the mistake in our reasoning? What have we failed to make use of in our proof?

instrument *A* to the left of the source will show the light to be moving at a velocity *c* − *v*. Reciprocally, instrument *B* will record the velocity of the light beam as *c* + *v*. In short, the velocity of the light source should be compounded geometrically with that of the light beam. Hence, the speed of light, which is always the same relative to its source, must, naturally, have different values with respect to the frame of reference in which the source is moving.

An exact analogy can be obtained by considering the observed effects when a gun-shell is exploding. All the scattering fragments will have the same speed relative to the centre of gravity of the shell.

A ballistic theory is enunciated.

Take two reference frames, one connected with the Earth, the other, with the centre of gravity of the set of fragments, placing the coordinate axes so that they coincide at the instant when the shell explodes. In time *t* the fragments will all be on the surface of a sphere of radius *ct* whose centre of gravity coincides with the centre of gravity of the fragments, and hence, with the origin of the coordinate system related to that centre. The centre of this sphere, however, will no longer coincide with the origin of the coordinate system of an

t- TIME
C-VELOCITY OF FRAGMENT
V-VELOCITY OF SHELL

observer on the ground. He will declare that
at time t the centre of the sphere made by the
fragments is located at a distance vt from the
origin. Furthermore, in the reference frame of
the centre of gravity of the shell, the speed of
its fragments does not depend on the direction
of their flight and is constant, while in the
reference frame "Earth" velocities do depend
on the direction, varying from $c - v$ to $c + v$.

Because of this analogy the theory according
to which the velocity of light depends on the
motion of the light source is sometimes called
the "ballistic" theory of light.

This theory, evidently, gives a simple
explanation to the negative result of the
Michelson experiment.[§] However, the deeper
the conclusions of the ballistic theory of light
were investigated the more hopeless the
picture grew, for it provided no satisfactory
explanation for such phenomena as refraction,
reflection, interference and diffraction of light.

More, in 1913 it was demonstrated that the
apparent motion of binary stars simply contradicts the ballistic theory. We are not concerned
with what the contradiction was and what
binary stars had to do with it. We shall accept
the fact that in 1913 experiments refuted the
ballistic theory.

Note that it is the year 1913, eight years
after Einstein had published his paper. When
Einstein was writing it there were no experiments directly contradicting the ballistic
theory of light. In fact, the theory itself was

[§]You will recall that the relativity principle was sufficient
to explain the Michelson experiment.

enunciated by Ritz only in 1908.[1] Undoubtedly, in the course of his work Einstein was bound to investigate the hypothesis, which was later propounded by Ritz. As a matter of fact, once the relativity principle is accepted, the ballistic hypothesis actually suggests itself, knocks on the door, so to say. Moreover, we have just sought to prove that it is the only possible conclusion to be drawn from the relativity principle. Go through our reasoning again and try and find a mistake! Everything seems flawless and logical. But if that is so, then we are in for some trouble. Here is where we are.

An impasse. An analysis of the situation.

1. Michelson's experiment leads us to the relativity principle.

2. By accepting the relativity principle we seem to imply that the velocity of light should depend on the velocity of its source, thereby establishing the ballistic theory.

3. We claim that many experiments refute the ballistic theory.

These three statements cannot be reconciled, which means that one of them is wrong. The relativity principle, it seems, is true. That leaves either the second or the third statement. Which one should we reject?

Today we know that the second statement is the wrong one, yet it had seemed most convincing. Let us review the reasoning which led to our conclusion: it seems logical enough. Remember also that Einstein could not be sure of the truth of the third statement.

The decisive experiments refuting the ballistic theory were made only in 1913 (De Sitter).

---

[1] Which means that by mentioning Ritz's theory before discussing the relativity theory we have violated the chronology of events.

At Einstein's disposal were only indirect arguments against the ballistic theory, arguments whose validity could be appreciated purely intuitively. Intuition, however, is an unreliable councellor. What we have so far is that

(i) the relativity principle and

(ii) the independence of the velocity of light on the motion of its source are irreconcilable!

A discomforting, but happily erroneous, conclusion.

If we accept the first, we must reject the second and accept the ballistic theory. If we accept the second we must give up the relativity principle and go back to the ether.

Einstein chose what could be regarded as the most impossible of all possible explanations of the Michelson experiment by taking these two propositions as the basic postulates for his theory.

It is difficult to comprehend how he arrived at this. A decisive part was probably played by that mysterious and elusive force called "intuition". Einstein himself says that from the time he was sixteen he had puzzled over the question: What will happen if a man tries to catch a light ray?

"A man moving with a ray of light at the velocity c (the velocity of light in vacuum) would see the light as an oscillating electromagnetic field, yet standing still in space. But nothing of the kind is possible. We know this both from experiment and from Maxwell's equations. From the beginning I seemed to feel intuitively that for such a man all phenomena should be the same as for the man who stands still relative to the Earth".[**]

[**]Incidentally, the paradox of the observer "astride" a light beam appears also in the ether theory and the

Another interesting, though hardly relevant, effect would be observed by a person travelling faster than light: he would see events reversed, just as we can reverse a motion picture film.

And so, Einstein failed to see how a satisfactory theory could be constructed on the assumption that the velocity of light differs from reference frame to reference frame and depends on the motion of the light source in a given frame. He puzzled this out for several years before he arrived at the conclusion: it was necessary to change — not merely electrodynamics — no, the whole of physics. Quite simple, you see.

Here is how he formulated the postulates for his new theory in his first work:

Attention! Einstein's postulates.

1. *The laws according to which the states of a physical system change do not depend on which of the two reference frames, in uniform relative motion, these laws refer to.*

2. *Every light ray moves in a "rest" reference frame* [in any given inertial frame – V.S.] *at a definite speed c, whether emitted from a stationary or a moving source.*

Thus, first, the relativity principle; second, the independence of the speed of light on the motion of its source.

And what about the ether in the new theory?

"The introduction of the 'luminiferous ether'," Einstein writes, "will prove to be superfluous

---

ballistic theory. The only difference is that in the ballistic theory we can investigate the paradox in any inertial frame. In the ether hypothesis we need an absolute frame, that is, the world ether. Einstein, we see, could not explain the paradox at first.

inasmuch as the view to be developed will not require an 'absolute stationary space' provided with special properties...." Einstein, we see, solves the ether mystery with a single master stroke: there is no ether. There is no hypothetical medium possessing mechanical properties. Space itself is capable of transmitting electromagnetic waves, that's all. In general, the relativity principle alone is enough to destroy the ether; even Ritz's ballistic theory did away with it.

The ether concept can now be scrapped, one way or the other. The relativity theory provides a ready explanation of the failure to detect the Earth's "ether drift"; even without Michelson's experiment the ether had caused a lot of trouble.

The second postulate, taken by itself, also seems quite acceptable (incidentally, it was accepted in the stationary ether theory).

Another attack on Einstein's postulates.

But how to reconcile the two? We have just shown that the relativity theory leads uniquely to the ballistic theory. If we recall the manner in which the dependence of the velocity of light on the motion of its source was proved, we can notice that, besides the relativity principle, only the formula for the composition of velocities, well-known in mechanics, was used. Here it is.

If the velocity of a body B relative to a body A is v_1, and the velocity of a body C relative to the body B is v_2, then the velocity of body C relative to body A is $v_1 + v_2$. Any doubts concerning the validity of this formula is tantamount to questioning the very fundamentals of mechanics.

Let us return to the example with the light flashes in a train carriage and see whether we can reconcile the relativity principle with the postulate of the constancy of the velocity of light.

We shall try to refute Einstein. We will use the classical method of *reductio ad absurdum* to expose the insolvency of relativity theory. Thus, we accept Einstein's postulates and see whither they lead us.[††]

Let an instantaneous flash of light have

[††]A minor digression is appropriate here. As in classical physics, the special theory of relativity treats "empty space" as "homogeneous and isotropic". In other words, the physical properties of space are identical in all directions. There is no favoured direction (isotropy), and every point in space is as good as the other (homogeneity). The isotropy and homogeneity of space are presumed throughout the whole of the following.

Strictly speaking, this is a kind of postulate stemming from experience. The isotropy of space with respect to electromagnetic processes is seen, for instance, in the fact that the velocity of light is the same in all directions (the front of a light wave is a spherical surface). Once again we note with satisfaction that physical declarations, even of the most general nature, are always prompted by experience alone. Without reference to experience, such declarations are meaningless. For all we know "empty space" might have turned out to resemble a crystal and be anisotropic. Only experience tells us the contrary, and it is experience that we believe in.

occurred precisely in the middle of our carriage. The light wave propagates in all directions, and in some infinitesimal time interval it reaches the end walls of the carriage. Suppose we can register this moment with very precise instruments. Will the flash reach the two walls simultaneously or not?

An observer using a reference frame rigidly connected with the carriage (he is inside the carriage, of course) definitely says, yes. The velocity of light is the same in all directions (the

isotropy of space), the distance to the walls is the same (the source is in the middle), and the carriage is at rest. Naturally, the flash reaches the walls at the same time.

An observer on the ground, however, whose reference frame is attached to the railway track, will claim that the flash reached the two walls at different instants. In his frame, to be sure, the light source is moving, but by Einstein's second postulate, this does not affect the propagation of light. The light signal will anyhow propagate in all directions at the same velocity relative to its source. But the forward wall of the carriage is moving away from the light signal and the opposite wall is approaching it. Hence, the signal reaches the latter a little earlier.

A strange situation with the concept of simultaneity of two events.

We find, thus, that two events that are simultaneous in one frame may not be simultaneous in another. A very strange result. Maybe this proves that Einstein's postulates contradict each other?

The number of such paradoxical examples could be cited endlessly. The foregoing, however, should be sufficient to make one wonder.

To sum up. From the outset the corollaries of his postulates confronted Einstein with the alternative: either his theory was insolvent, or all the fundamental concepts of space and time, the whole structure of physics as a science, must be revised.

At once the new theory overshadows the important but very special, from the point of view of new horizons, question of the "electro-dynamics of moving media". The question now is not merely of destroying the ether (which now seems paltry indeed) but of nothing less than a critical analysis and reappraisal of the fundamentals of physics.

I repeat that to me personally it remains a mystery how Einstein ever risked expounding his theory. It is often claimed that, before establishing the principles, he chose from the mass of experimental data only the absolutely reliable facts, which is why he was absolutely confident of his postulates. This is but partially true. Experiments did lead him to the relativity principle. But the principle of the constancy of the velocity of light had no direct confirmation. More, we have seen that the price to be paid for the simultaneous acceptance of relativity and the constancy of the velocity of light was exorbitant.

The most natural way out was probably suggested by Ritz in his ballistic theory, according to which the ether was discarded and the Maxwell equations altered. This, of course, was also revolutionary, but compared with the relativity theory it looks peaceable and patriarchal.

Einstein completely revises such physical fundamentals as time, space, simultaneity, the Galileo transformation and interaction at a distance. In effect, what he demonstrates is that there are no fundamentals, that all these "apparently familiar concepts" are simply non-existent in classical physics, that physicists had generalised experience intuitively and unconsciously without stopping to take a look at what they were doing and what their theories were based upon. And all this, in the final analysis, he did solely on the basis of Michelson's experiment. I find that there is no other such example of intellectual courage in the whole history of science.

And finally, extremely rigid demands are naturally imposed on the relativity theory. This theory, which calls for entirely new and at first glance paradoxical concepts of space and time, will not be forgiven nebulous hypotheses,

Non-Euclidean geometry alone can, possibly, claim a place alongside of Einstein's work.

approximate explanations or logical flaws. Such a theory must be crystal clear and logically flawless. All known experimental facts without exception must be clearly and unequivocally explained. And lastly, all physical laws, the validity of which has been confirmed experimentally, all unchallenged physical theories must remain within certain classes of phenomena approximate and valid to a high degree of accuracy. The new theory may generalise, but it cannot reject Newtonian mechanics.

Einstein had to meet all these requirements in the very first stage of his work, and this he did. The least that can be said is that it is incredible how twenty-six-year-old Einstein could have ever produced his theory.

And now, for the time being, let us forget all that has been said before. The postulates of the special theory of relativity are in fact very natural. We shall now endeavour to investigate this theory.

CHAPTER XII,

which expounds in considerable detail on the postulate of the uniformity of the velocity of light and then goes over to discuss the concepts of time and simultaneity in relativity theory

EINSTEIN (simultaneity, time)

It will probably be best to start with a remark on terminology. Several times in the foregoing discourse the "point of view of the observer" was mentioned. In the previous chapter, in analysing the relativity of the concept of simultaneity, we had "the point of view of an observer inside a train carriage" and "the point of view of an observer on the ground". This is accepted physical terminology, and it was introduced by Einstein. It is graphic and convenient, and we shall make use of it in the future. Still, it has one drawback, for once we speak of an "observer" and his "point of view" we begin to suspect, on a purely linguistic basis, that physicists must be standing on subjective platforms.

A brief digression.

This, of course, is not the case. What is meant is not the personal, subjective perception of the observer but concrete physical measurements carried out in specific physical conditions. The "observer's point of view" appears whenever relative physical notions are examined. We say, for example: "In a frame of reference rigidly attached to a train carriage the speed of the carriage is zero." When we state that the velocity of the carriage is zero "from the point of view of an observer inside the carriage" we are saying the same thing in different words.

Velocity is a relative physical concept. The determination of velocity or, broader, the technique for measuring velocity, is inseparable from the concept of frame of reference. In a given frame of reference the velocity of a body is measured by certain purely objective means.

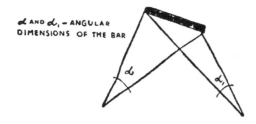

α AND α_1 – ANGULAR DIMENSIONS OF THE BAR

It should be emphasised that there is not a single subjective concept, definition, or quantity in the whole realm of physics.

All this holds equally with respect to other relative physical quantities and concepts. An example is the angular dimensions of an object. You would hardly claim that the angle at which an object is seen is a subjective category. True, angular dimensions depend on the distance from a body and the angle at which it is observed. But they are measured by absolutely objective means. And just as the angular dimensions of a sphere change depending on the distance of the measuring instrument from it, "velocity", "simultaneity" and other values change depending on the frame of reference in which the measuring instrument is located.

To begin with, let us alter the second postulate slightly. The change will first be of a purely verbal nature, but the principle of the independence of the velocity of light on the motion of its source will be visualised clearer.

"The velocity with which the front of any light wave travels in any inertial reference frame is the same in all directions."*

Now for a more essential change. "The maximum velocity of transmission of any action (signal) at a distance in any inertial frame is finite and does not depend on the motion of the source of the action."†

Getting back to the promised systematic analysis of the fundamentals of the special theory of relativity.

Postulating the finiteness of the maximum velocity of transmission of any signal at a distance.

---

*A minor piece of logical subtlety! The fact that the speed of a light wave does not depend on the motion of the light source relative to a given reference frame is taken into account in our definition, which speaks of "any" light wave (including waves emitted by moving sources). The remark concerning the constancy of the speed in any direction implies the isotropy of space. By the relativity principle, our definition is valid in any inertial frame if it is valid in one.

†In parentheses we have the word "signal". By signal nothing is meant which contradicts to everyday notions. A signal is anything that can conceivably be used to transmit to a point A information about doings at a point B, i.e., a signal is a vehicle for transmitting information.

It should be clear why the term "action" is equated to the term "signal". A signal can be transmitted only by some action (electromagnetic, gravitational, etc.) which takes place between points A and B. The maximum velocity of transmission of this action is the velocity with which the signal (information) can be transmitted. Note,

We shall have to take it for granted that an analysis of Einstein's postulates leads to the conclusion that there exists a maximum velocity of transmission of any action (signal), and that it is equal to the speed of light in vacuum. We see that the principle of the uniformity of the speed of light is part of a much more general law of nature.

Generally speaking, the proposition that there exists some maximum velocity with which an action can be transmitted sounds more natural than the hypothesis of classical physics which implies that some actions can be transmitted with infinite speed.

Simultaneity.

We shall not investigate the manner in which the principle of the uniformity of the velocity of light leads to the more general principle of the finiteness of the velocity of transmission of action (or information) at a distance. We shall accept this as true and proceed to analyse one of the basic conceptions of relativity theory, that of the simultaneity of two events. Let us first recall how the problem was tackled in classical physics. Before Einstein, no one stopped to analyse the concept of simultaneity. It was regarded as obvious. Classical physics, of course, had a very definite notion of what was meant by the simultaneity of events taking place at a distance, but it was a subconscious, intuitive generalisation of experience.

In Chapter III we established the approach of classical physics to simultaneity:

Recalling the classical definition of simultaneity.

Two events, such that, generally speaking, either one could be the cause or the effect of the

finally, that in relativity theory the concept "event" is primary and its meaning corresponds to the ordinary, lay meaning of the word.

other, can be simultaneous only if neither is the cause nor the effect of the other.

It is apparent that the classical conception of simultaneity was prompted by the principle of cause and effect. Less obvious is the fact that this definition tacitly promotes a most important hypothesis: The maximum velocity of transmission of an action (the speed of transmission of information) is infinite.

Later on we shall see that only in this case is our definition unique. As was previously mentioned, if we accept that information can

be transmitted with infinite speed, we arrive at the conclusion that two events simultaneous in one frame of reference are simultaneous in any other. In this sense the simultaneity of events is, in classical physics, an absolute concept.[‡]

In pre-Einsteinian physics there even existed a model of action transmitted with infinite speed, the absolutely rigid body. A rigid body moves through space as an entity, and therefore it transmits information instantaneously. If you push point A of such a body from rest, point B will start moving at the very same instant. People did not suspect that the concept of a rigid body was in principle an impermissible

---

[‡]Absolute simultaneity leads directly to the absolute time of classical physics, but we will not go into this aspect of the matter.

idealisation of real physical bodies. This was realised only after Einstein.[§]

But as soon as we reject the existence of signals possessing infinite velocity, the classical notion of the simultaneity of two events at different points in space becomes insolvent. Let us demonstrate this.

Assume, first, that an action is transmitted with finite velocity, and secondly, that this velocity is the same in any inertial frame and is equal to the speed with which the front of a light wave propagates in vacuum. Consider two events A and B, using for convenience a specified

---

[§]An interesting point, though not quite following from the foregoing, is that extreme care must be displayed in operating, in classical physics, with the declarations: "two non-simultaneous events at one point" and "two non-simultaneous events at different points". These take on some meaning only when a frame of reference is indicated. A passenger of the Moscow–Leningrad express using a frame of reference attached to the train can claim that two such events as the engine's whistle on leaving Moscow and arriving at Leningrad took place at the same point in space. It hardly needs to be proved that to any person on Earth these two events happened at different points. Thus, the coincidence of two non-simultaneous events in space is relative even in classical physics.

But the coincidence of two events in time is, in classical physics, an absolute conception. That is to say, if two events are simultaneous in one frame of reference they are simultaneous in any other.

The absoluteness of the coincidence, or non-coincidence, in space of two simultaneous events is postulated in classical physics, though it is hard to discuss whether it is due to the infinite velocity of transmission of actions or not. We are little concerned with such niceties. Strictly speaking, the absoluteness of simultaneity is also postulated and does not follow directly from the hypothesis of the infinity of the velocity of transmission of action at a distance.

frame of reference. Also for convenience, assume that the physical manifestations of events A and B are absolutely identical, say, identical flashes of light (this condition is needed only to clarify the meaning of what follows and does not affect the principle).

Let flash A have taken place at time t_A at a point given by the coordinates x_A; y_A; z_A and flash B at time t_B at a point x_B; y_B; z_B. The distance between the two points is

$$r_{AB} = \sqrt{\left(x_B - x_A\right)^2 + \left(y_B - y_A\right)^2 + \left(z_B - z_A\right)^2}.$$

To simplify our analysis, suppose that in our frame of reference flash A preceded flash B, and consequently t_B is greater than t_A. The minimum time for the news of event A to travel from point (x_A, y_A, z_A) to point (x_B, y_B, z_B) is

$$t_{inf} = \frac{r_{AB}}{c}.$$

If $t_B - t_A > t_{inf}$, then in principle it is possible for a person at point (x_B, y_B, z_B) to receive the news of flash A before flash B takes place. In this case events A and B may be linked causally, and flash B may be regarded as an effect of event A.

If, conversely, $t_B - t_A < t_{inf}$ then, *in principle*, any news about A will reach point (x_B, y_B, z_B) after flash B has occurred, and no causal relationship between events A and B is possible. Hence, event B cannot be an effect of event A.

Thus, if information is transmitted with finite speed, then to any given event A at a point (x_A, y_A, z_A) there can be an infinite number of events B_1, B_2, B_3, ... taking place at different

Several precisely formulated concepts.

instants at points (x_B, y_B, z_B) which can have no causal links with A.

All this is quite lucid and the only inconvenience is the abstract nature of the discourse. Let us consider an illustration.

Poets like to wonder about the fact that many of the stars in the sky may have been extinct for ages. If at the moment when you are reading this line, a gigantic explosion occurs on the nearest star (Proxima Centauri), reducing it to smithereens, we will see the

flash heralding the catastrophe only four years hence. Our world works in such a way that there is no conceivable means of learning of the event before a beam of light brings us the news. The star will have been dead for more than four years, and all this time we will be seeing it twinkling placidly in the sky.

When astronomers record the outburst of a nova in some distant galaxy, it means that the catastrophe took place long before our monkey forebears ever contemplated leaving their tree dwellings to start evolving towards man. Nothing can bring us the news any faster. So says Einstein.

Thus, event A — the explosion of Proxima — and event B — photographing the outburst on Earth — are separated by a time interval t_{inf} of four years. These two events are linked by a relation of cause and effect and in principle cannot have been simultaneous.

But any other pair of events taking place on Proxima and on Earth and separated by a period of less than four years can, in principle, be simultaneous, for they can have no causal relation!

We have thus naturally arrived at the need to change and clarify the meaning of simultaneity, a concept which is absolutely essential. Without it we could not compare time (clocks) at different points, it would be impossible to measure the length of moving bodies, in short, it would be impossible to develop physics as a science. Accordingly, Einstein commences the exposition of his theory with a definition of simultaneity.

Two events, taking place at points A and B of an inertial reference frame O, are simultaneous if the light signals (or any other signals of finite velocity) sent from points A and B at the instants of the events reach a point at the middle of line AB at the same time.

Two definitions which constitute the sum and substance of this chapter. All else is "garnish". It would now be worth having another look at Chapter III.

Now we have everything necessary for a precise definition of the concept of time. It is defined as *the sum total of readings of identical clocks placed at different points of space within a frame O, at rest in that frame and giving the same readings simultaneously.*[¶]

You see that the conception of simultaneous events has changed markedly since Chapter III. The new definition of simultaneity, like all physical notions, is forced upon us by the realities of life and reflects the objective world. Of course, no end of concepts can be introduced and defined, but physics has no place for conceptions which fail to reflect the real world, and physicists do not care for them.[**]

The likeness and difference between the definitions of simultaneity in Newtonian and Einsteinian physics.

Einstein's notion of simultaneity of events at different points of space stems naturally from the sum total of experience. Like the "classics", Einstein bases the concept of simultaneity on the principle of causality. In nature there is no other image of the simultaneity of two events than the impossibility of any causal relations between them. Without going into other subtleties associated with the concept of simultaneity (and there are no end of them)

[¶]Incidentally, without the concept of simultaneity the whole of the foregoing analysis would, strictly speaking, have been meaningless, since we had no idea of time in different points of space. This inaccuracy, however, was intentional, insofar as the need for providing a new definition of simultaneity is obvious from the previous discourse, and to make it flawless several minor clarifications are sufficient.

[**]Incidentally, in mathematics scientists often operate with schemes that have no direct relation to reality. Recall non-Euclidean geometry. Many geometries can be devised, only one is clearly manifest in the real world.

we will point out the fundamental difference between Einstein's conception of simultaneity and classical views.

As long as we speak of simultaneous events at one point, everything is clear and relativity theory conforms with classical notions. But as soon as we begin to analyse what is meant by two simultaneous events in different places, the situation changes.

Following Einstein's reasoning, we accept that the maximum velocity of transmission of information is finite, though very great (300,000 km/sec). Thus, for the signal of event A to reach point $(x_B; y_B; z_B)$, a time interval is necessary. Furthermore, any event B_1, B_2, B_3 taking place at point $(x_B; y_B; z_B)$ at some instant within that interval can, *in principle*, in no way be causally linked with event A. Which of these events is simultaneous with event A? Can they all be considered simultaneous? This is obviously ill-advised, so we have to find one event B which can be declared simultaneous with A. Which one?

The answer seems to suggest itself: "Why of course the one that took place at the same instant of time." Even though we realise that this phrase carries no meaning, we want to say it, so obvious it seems. It is very difficult to get rid of the convention that time is something absolute; independent and so self-evident as to make any discussion of it unnecessary. Probably the most difficult barrier to the understanding of Einstein's theory is the difficulty of discarding commonplace, conventional notions.

But to get back to our problem. Which of the infinite number of events B_1, B_2, B_3, ..., which cannot be causally linked with event A, can be declared simultaneous with the latter? Einstein advances the only possible definition of simultaneity under the circumstances. Why is it "the only possible one"?

Note the definition of synchronised clocks given here.

It is easy to see that, according to Einstein, the definition of simultaneity leads to the following procedure for verifying the synchronised motion of clocks (two clocks are said to be synchronised if they simultaneously show the same time). Let there be two clocks at points A and B of an inertial frame O. At time t_{1A} (3 o'clock on clock A in the drawing) a light signal is beamed to point B, where it is instantaneously reflected and returns to point A at time t_{2A} (9 o'clock according to clock A).

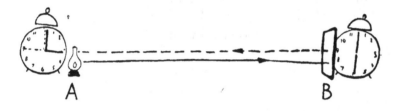

Now, if at the instant when the signal reached point B the time by clock B was

$$t_B = \frac{t_{1A} + t_{2A}}{2}$$

(6 o'clock on clock B), then clocks A and B are synchronised.

It can be demonstrated that if we accepted some other definition of "simultaneity" than Einstein's, clock B would have to be regarded as synchronised with clock A only if at the time of the arrival of the light signal at B it showed, say, not 6:00 but 6:15 or 5:45. But this would contradict our notion of the isotropy of space. For, on the one hand, we assume that the beam of light takes the same time to travel from A to B as from B to A. On the other hand, in determining the time it takes the beam to travel back and forth by the synchronised clocks B and A, we would find that it took the beam 3 hours and 15 minutes from A to B, and 2 hours 45 minutes back from B to A. Which is why Einstein's definition is the only possible one.

We have dwelt on these subtleties not only because simultaneity is a fundamental concept of relativity theory. It must be said that even such a very fragmentary analysis, interspersed as it is with endless requests to take this or that assertion for granted, even such an analysis, which hardly deserves its name, is rather tiring. But the best way to demonstrate what real science is, is to see its inner workings. Then one can gain an idea of the mental labours required to give birth to conclusions which later become so conventional as to seem commonplace. In the same way we might attempt to describe how much strength, stamina, energy and courage a boxer in the ring needs, but a person who has ever tried even shadow boxing for a few minutes will realise this much better.

To return to our "ring", the reader is warned that he must prepare for another telling blow. We shall soon see that the simultaneity of events is a relative concept, that events simultaneous in one inertial frame may not be simultaneous in another. But first one more general remark.

Again the author soliloquises.

Einstein is not to blame for making simultaneity and time relative concepts. That is just how the world is made. Before Einstein no one suspected that such concepts as simultaneity, time or length required cast-iron definitions. That is why relativity theory was no less a psychological than a physical revolution. The situation was aptly characterised by L. M. Mandelshtam:

"That the conception of simultaneity had, as Einstein stressed, to be defined and was not ordained from above was a step which no one can retract."

It must be clearly understood that neither the postulates nor the basic concepts of Einstein's theory come into contradiction with ordinary logic. The theory may contradict some facts: that is another question. But so far experimental data from the most different fields brilliantly confirm it.

The relativity of simultaneity is one of the focal points of relativity theory.

Let us go back to the example discussed in the previous chapter. A flash of light occurs in the middle of a carriage in uniform motion. In the reference frame attached to the train the light signals reach the end walls simultaneously. In the frame attached to the ground the two events are not simultaneous. This "strange" conclusion is correct, and the example clearly illustrates the relativity of simultaneity.

But here is an interesting point. It may not have occurred to you that until a clear-cut definition of simultaneity was formulated (whether according to classical mechanics or to Einstein) our discourse was meaningless.

The declaration that two events are simultaneous or otherwise makes sense only when the concept of simultaneity is defined. This is not an ordained or an *a priori* concept. We have to formulate it ourselves and, I repeat, this formula is forced upon us by the real world surrounding us.

Many examples illustrating the relative character of simultaneity could be cited, but we shall restrict ourselves to the example cited by Einstein himself.

A train is travelling on a railway.[††] Lightning strikes the ends of the train and passes into the ground. We want to establish whether the lightning bolts were simultaneous or not. An

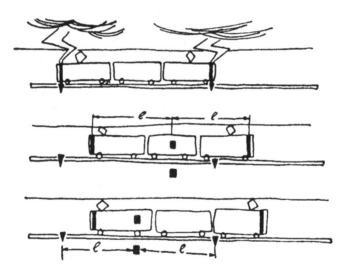

observer in the train will declare that the bolts were simultaneous if an instrument (say a photoelectric cell) located *exactly in the middle*

[††]To visualise the experiment better, imagine that the train is 10 light years long and is travelling at 150,000 km/sec.

A classical example il-
lustrating the relative
character of simul-
taneity. Incidentally, it
is rather difficult.

of the train registers the two light signals at precisely the same instant.

An observer on the ground will declare that the bolts were simultaneous if the light signals were registered at the same instant by an instrument located *exactly halfway between the marks left by the lightning on the ground*. We are not concerned with the technique of the experiment, but to escape possible misunderstandings we assume that there are two sets of instruments: in the train and on the railway embankment. Furthermore, the instruments are triggered only when the two light signals reach them at the same time (the instruments are provided with coincidence circuits). When the lightning has struck, we check both set-ups to see which one was triggered. If it is the one in the train then, by definition, the lightning bolts were simultaneous in the train frame. In the set-up on the ground, apparently, the instrument precisely opposite the middle of the train will be triggered. But the light signals took some time to reach the instruments, and the train has travelled some distance. Its middle is not at the halfway point between the lightning marks on the track but closer to the "front" mark! Therefore, in the ground reference frame the lightning bolts, according to our definition, are not simultaneous, and the ground observer will say that the rear of the train was hit first. In the same way, if it is found that the instrument standing halfway between the lightning marks on the ground was triggered, in the train it will be observed that an instrument closer to the rear was triggered. In this case the lightning bolts were simultaneous in the ground frame and not simultaneous in the train frame. And there is no conceivable way in which the two bolts can be simultaneous in both frames at once.

Why have we dwelt in such detail on the concept of simultaneity? There are at least two reasons.

First, the simultaneity concept is fundamental in Einstein's theory. If it is understood clearly then the physical principles of relativity theory become natural and clear. That is why Einstein begins his theory with the concept of simultaneity.

We could, of course, offer a "circumspect" exposition of relativity and hustle simultaneity in through the back door, without defining it openly, and introducing only the notion of the synchronised clocks. This, however, would hardly have been honest and, furthermore, it would obscure the core of the matter.

The second reason why simultaneity had to be discussed is that the concept in Einstein's theory aroused a considerable controversy. Some authors, without taking pains to investigate the problem, hastened to declare Einstein's exposition of simultaneity as being contrary to dialectical materialism. Depending on their own views, people hail or reject Einstein's notions of the physical framework of his theory and, in particular, of the simultaneity concept.

Remarks on the problem from the philosophical point of view.

Speaking of simultaneity, we should mention its philosophical aspect, though the author realises only too clearly how incompetent he is in this sphere. All that has been said about simultaneity serves to illustrate once again the justification of the methods of materialistic philosophy. To the materialist it is obvious that in physics there is no place for *a priori* conceptions. Hence, the concept of simultaneity must be defined, and experience leads us to the content of this concept. The relative nature of simultaneity, and accordingly time, does not disturb the materialist. The materialist does not force his notions on nature. On the contrary, the exploration of real life leads the scientist to his formulation of the various conceptions reflecting reality.

And that's about all. No end of scrutinising can detect any contradiction between simultaneity as considered by Einstein and the principles of dialectical materialism.

In conclusion, a general remark. The importance of Einstein's theory as a method lies, first and foremost, in that it has demonstrated that some concepts postulated in science are quite meaningless (such as Newton's "absolute space"). The reciprocal of this is the wide use of "self-evident", *a priori* notions, such as simultaneity, length and time in classical physics.

It would seem that after Einstein there could be no place for such views in physics. Yet, paradoxically, the main arguments raging about the interpretation of relativity theory are due precisely to a hasty use of words without a clear understanding of their meaning.

CHAPTER XIII,

which informs the reader in a very matter-of-fact way what is meant by "spacetime interval" and the Lorentz transformation. Towards the end of the chapter, if he ever gets to it, the reader will find out the curious formula for adding velocities in Einstein's theory

EINSTEIN ("astonishing" conclusions)

At the risk of over-simplification, it can be stated that the mathematical concept of Einstein's theory is based on one fact: the invariance of spacetime interval. What is meant by "spacetime interval" and its "invariance" will be explained forthwith. True, in our discourse the meaning of the interval concept will not be explained. Thus, when the author assures the reader that it is a very important notion he resembles a person displaying a photograph of a tiger as proof of the beast's ferocity. The unsophisticated interlocuter, however, has a lingering impression that he was shown merely an enlarged picture of a cute kitten. Nevertheless, the speaker is unable to resist the temptation of displaying the photograph.

274

Let two events A and B have taken place at points x_A, y_A, z_A and x_B, y_B, z_B in some specified inertial reference frame O, and, furthermore, let the time of these events as measured in that frame be t_A and t_B respectively. The spacetime interval between the two events is given by the equation

$$S_{AB}^2 = c^2 \left(t_B - t_A\right)^2 - \left(x_B - x_A\right)^2$$
$$- \left(y_B - y_A\right)^2 - \left(z_B - z_A\right)^2.$$

This quantity possesses an extraordinary property.

Suppose our events A and B were observed from another inertial reference frame O', the space-time coordinates of the events in the new frame will be x_A', y_A', z_A', t_A' and x_B', y_B', z_B', t_B'. Once again we shall visualise our case with the help of our old friend the railway, such that the reference frame connected with the track is inertial. Suppose this is our unprimed frame O. (Remembering that, strictly speaking, a reference frame connected with the Earth is not inertial, we shall have to lay our track somewhere in outer space.)

Suppose a train goes with uniform speed along our straight track. This means that the primed reference frame O' connected with it is also inertial. Let now somewhere in the heavens two super-novae explode: events A and B.

Now, if the observers on the track and in the train register the space-time coordinates of the events they will find that

$$S_{AB} = S_{AB}', \text{ or } c^2 \left(t_B - t_A\right)^2 - \left(x_B - x_A\right)^2$$
$$- \left(y_B - y_A\right)^2 - \left(z_B - z_A\right)^2 = c^2 \left(t_B' - t_A'\right)^2$$
$$- \left(x_B' - x_A'\right)^2 - \left(y_B' - y_A'\right)^2 - \left(z_B' - z_A'\right)^2.$$

The invariance of spacetime interval and a sprinkling of mathematics.

This is what is meant
by the invariance of a
spacetime interval.

The interval between the two events is
the same in two different inertial frames, or
in other words, it is invariant. The foregoing
equation can be more conveniently written
down in the form

$$S_{AB}^2 = c^2 t_{AB}^2 - r_{AB}^2 = c^2 \left(t'_{AB} \right)^2 - \left(r'_{AB} \right)^2 = \left(S'_{AB} \right)^2,$$

where r_{AB} and r'_{AB} give the space distances, and
t_{AB} and t'_{AB} the time intervals between events

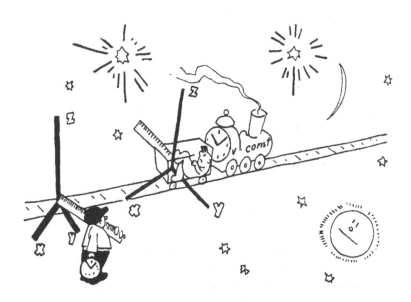

A and B in the respective reference frames.

How was it established that a spacetime
interval does not change in passing from one
frame to another? The invariance of spacetime
interval represents simply the mathematical
dress of the basic propositions of the theory:

the relativity principle plus the principle of the uniformity of the velocity of light. There is no sense in going into the actual proof, even though it is fairly simple. This is a mathematical problem, and mathematics, as the late Academician A. N. Krylov liked to say, is like a mill which grinds everything that falls between the millstones. Our interest lies in the "grist".

Sprouting directly from the invariance of spacetime interval is the so-called Lorentz transformation, which provides a formula for going over from one inertial reference frame to another. This, too, belongs to the realm

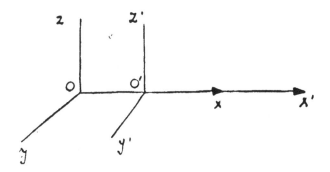

of mathematics. We shall not go into the mathematical proof of the Lorentz transformation; we shall even, very reluctantly, refrain from presenting the extremely subtle mathematical explanations of the transformation developed by Minkowski. All this also belongs to the functions of our mill. As it is, we shall have our hands full wading through the basic physical conclusions of the theory. We shall therefore take all formulas for granted.

These equations give the Lorentz transformation.

Consider two inertial reference frames O and O' whose axes coincide. Let the relative velocity v of the frames be directed along the axes x and x'. Then, knowing the space-time coordinates of any event in one frame, we can find the respective coordinates in the other frame:

$$x = \frac{x' + vt'}{\sqrt{1 - \dfrac{v^2}{c^2}}},$$

$$y = y',$$

$$z = z',$$

$$t = \frac{t' + \dfrac{v}{c^2}x'}{\sqrt{1 - \dfrac{v^2}{c^2}}}.$$

These are the formulas for passing from the unprimed to the primed frame.[*]

From the drawing we see that in the case under consideration the velocity of frame O' in frame O is $+v$. Now, knowing the space-time coordinates in the primed frame, we can use our formulas to determine the respective coordinates in the unprimed frame. In order to make this reverse transition we must solve our equations with respect to x' and t' ("isolate" x' and t'). This is readily accomplished formally, but it is even simpler to recall that the equality

---

[*]It is worth noting that the formulas of the Lorentz transformation make sense only if the relative velocity of the frames is $v < c$. At $v > c$, it will be readily observed, the root in the denominator is imaginary. This, of course, could have been anticipated, since mathematical formalism must correspond to physical propositions, and you remember that velocities greater than c are impossible.

of the two inertial frames means that the formulas for passing from O to O' and from O' to O should be of the same form. Taking into account that the velocity of O relative to O' is $-v$, we can write immediately:

$$x' = \frac{x - vt}{\sqrt{1 - \dfrac{v^2}{c^2}}},$$

$$y' = y,$$
$$z' = z,$$

$$t' = \frac{t - \dfrac{v}{c^2} x'}{\sqrt{1 - \dfrac{v^2}{c^2}}}.$$

We have examined the fairly simple case when the relative velocity of the two frames O and O' coincides in direction with the x and x' axes. In the most general case, the transformation formulas are, naturally, more involved, but the essential differences between Einstein's theory and classical physics are all present in the special case.

The difference between the Lorentz transformation and the Galileo transformation, its counterpart in classical mechanics, is apparent. At the same time, though, they have much in common.

In this connection a very general declaration can be made. It is immediately obvious that Einstein's theory should cover classical mechanics as a limiting case. Newtonian mechanics has been repeatedly confirmed by experience, and no reasonable new theory can merely push it out of the way. Newton's method of principles is an eternal guarantee against any such mishap. Principles may change but, whatever

Newtonian mechanics as a limiting case. An important general remark illustrated by a concrete example.

may be discovered in the future, for bodies travelling at small velocities any theory must produce the same, or to be more precise, almost the same results as Newtonian mechanics. Newton's laws will always remain valid as a fair approximation of the truth.

Everything just said with regard to Newtonian mechanics could be repeated word

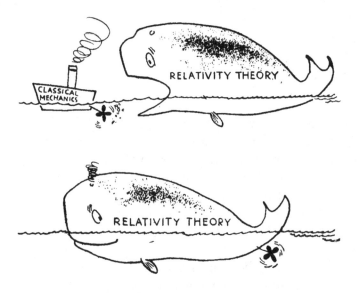

for word with respect to the special theory of relativity. The further progress of science may bring in any number of changes. Anything can happen, but Einstein's theory will always remain a fair approximation of the truth.

But to return to our question. We want to demonstrate how Einstein's theory incorporates Newtonian mechanics. This is readily achieved if we analyse some conclusion of the theory. One example will suffice. When $\dfrac{v}{c} \ll 1$, members of

the order $\left(\dfrac{v}{c}\right)^2$ and $\dfrac{v}{c^2}$ can be neglected, and

the Lorentz transformation formulas turn into the familiar formulas of the classical Galileo transformation:

$$x = x' + vt',$$
$$y = y', \quad z = z', \quad t = t'.$$

On the other hand, the Lorentz transformation develops into the Galileo transformation if c tends to infinity. Here the physical interpretation is also apparent: the propagation of signals with infinite velocity is the assumption, you will recall, underlying classical physics.

And now for a minor sensation. Our work is nearing completion. The whole of the special relativity theory stems immediately from the two postulates we investigated before. The main difference with classical physics lies in the concept of time or, what is the same thing, the concept of simultaneity. This, too, has been examined. We have not mentioned one extremely important conclusion of principle: the connection between mass and energy, but this will come.

As the mathematical expression of the theory is based completely on the Lorentz transformation, which we have examined, everything else, including length contraction and time dilution, are nothing more than corollaries. So let us consider some technicalities with a glowing sense of knowledge of the fundamentals. First, the law of composition (or addition) of velocities.

The statement of the problem is clear. Let a body be moving in some inertial reference

One of the most surprising conclusions of relativity theory for a person brought up on Newtonian mechanics — the law of composition of velocities.

frame O with a velocity v_1, and let another body be moving with a velocity v_2 relative to the first body. To determine the velocity of the second body with respect to the reference frame O. After the satisfaction of an accurate statement of the problem, let us go back to our railway.

Our train is moving with a velocity v_1 *relative to the track* (and there is nothing to prevent its velocity from approaching the speed of light). For a reason that does not concern us, one of the passengers shoots a rifle, and the bullet travels with a velocity v_2 *relative to the train*. What is the velocity of the bullet relatively to the track? (The bullet, of course, can also approach the speed of light.) Even though we restrict ourselves to the special case of the velocities v_1 and v_2 being in the same direction, the example brings out all the features of relativity theory.

In classical mechanics, the net velocity is given by the very simple formula $v_{net} = v_1 \pm v_2$ (the "+" sign denoting a bullet shot in the direction of the train's motion, the "–" sign, in the opposite direction).

According to Einstein, the law for determining the net velocity is different:

$$v_{net} = \frac{v_1 \pm v_2}{1 \pm \dfrac{v_1 v_2}{c^2}}.$$

We see that, at $v_1 \ll c$ and $v_2 \ll c$, Einstein's formula turns into the classical one, for in this case we can easily neglect the second term of the denominator as compared with unity. The change comes when the velocities v_1 and v_2 are comparable with that of light. This is best demonstrated if we take one of the velocities (v_2, for example) equal to that of light. If you remember, we had a similar problem when we discussed, in Chapter XI, the velocity relative to the track of a beam of light emitted by a source in a moving train. It is obvious that, *regardless of v_1*, the absolute net velocity equals that of light:

$$\frac{v_1 \pm c}{1 \pm \dfrac{v_1 c}{c^2}} = \pm c.$$

Now we are in a position to refute the reasoning of Chapter XI. You recall that in defending the ballistic hypothesis we took the classical formula for adding velocities as self-evident. And therein, we find, lay our mistake.

It would be worth re-reading page 248 now.

A light wave from the locomotive's headlight travels with a velocity c relative to the train. Its velocity relative to an observer on the ground is also c, not $v_{train} + c$.

To people brought up in terms of classical mechanics this seems incredible, and yet it is true. More, the relative velocity of two photons hurtling head-on, each with the speed of light, is again c, and not $2c$ as classical mechanics would declare.[†]

In relativistic mechanics the velocity of light in vacuum is an insurmountable barrier.

[†] It can be demonstrated easily that the problem of determining the relative velocity of two bodies is identical to that of determining the law for the composition of velocities.

CHAPTER XIV,

which discusses two corollaries of relativity theory which are usually the cause of much bewilderment

EINSTEIN (time, length)

In Chapter III we discussed how to measure the length of a moving body and came to the conclusion that "the length of a moving body is the distance between two simultaneously made intercepts of its initial and terminal points".

In classical physics, the length of a moving body so measured was the same as the length of the body at rest, and all was fine. Attention is again drawn to the following.

1. Before Einstein, no one stopped to consider "how" the length of a moving body was measured. Actually, in measuring, or speaking of, length the above definition was tacitly accepted.

2. The coincidence or non-coincidence of the length of a resting and a moving body is a question of experiment, and we cannot declare *a priori* that they should necessarily coincide.

We should never force our views and desires on nature. Within the specified reference frame where the measurements are made, a stationary bar and a moving bar are in different physical conditions; therefore we have no cause to expect in advance that the length must remain the same. Before, on the basis of everyday experience it was taken for granted that length does not change with motion. In conventional experiments it is impossible to observe any difference between the length of a body when it is at rest and in motion, as the attainable velocities of material bodies are immeasurably below the speed of light. That is why no changes in length were ever observed, and from this stemmed the conviction that the length of a body is absolute and constant regardless of the reference frame in which it is measured.

But a direct analysis of the Lorentz transformation immediately shows up length to be a relative quantity. The length of a bar moving with a velocity v contracts in the direction of the motion and is determined by the formula

$$l = l_0 \sqrt{1 - \frac{v^2}{c^2}},$$

where l_0 is the length of the bar at rest[*] (i.e., measured in the reference frame in which it is at rest).

---

[*]This equation is so easily derived that the procedure can be described here. In order to determine the length of a moving bar, an observer must make simultaneous

This is called the Lorentz contraction.†

For a space rocket travelling around the Sun, the contraction, as observed from Earth, is

$$l = l_0 \sqrt{1 - \left(\frac{11 \, km \, / \, sec}{3 \times 10^5 \, km \, / \, sec} \right)^2}$$

$$\approx l_0 \left(1 - 0.067 \times 10^{-8} \right),$$

or about seven one-hundred-millionths of one percent!

---

intercepts of its initial and terminal points x_1 and x_2. Then $(x_2 - x_1)$ is the length l of the bar.

To find the relation between l and l_0 through the Lorentz transformation, we must determine the relation between the coordinates x_1 and x_2 and the corresponding coordinates x_1' and x_2' of the initial and terminal points in the reference frame in which the bar is at rest:

$$x_1' = \frac{x_1 - vt}{\sqrt{1 - \frac{v^2}{c^2}}},$$

$$x_2' = \frac{x_2 - vt}{\sqrt{1 - \frac{v^2}{c^2}}}.$$

Note that the time t is the same in both right-hand terms. This is a reflection of the fact that the intercepts of the initial and terminal points were made simultaneously. Subtracting the lower equation from the upper one, we obtain

$$x_2' - x_1' = \frac{x_2 - x_1}{\sqrt{1 - \frac{v^2}{c^2}}}.$$

But $x_2' - x_1' = l_0$, and $x_2 - x_1 = l$, hence

$$l = l_0 \sqrt{1 - \frac{v^2}{c^2}}.$$

†The Lorentz contraction derives its name from Lorentz's hypothesis (mentioned in Chapter XI), which postulates

Obviously, there is no way of detecting such a contraction for so far space vehicles are the unchallenged speed champions of the macrocosm. It is not surprising, therefore, that length has always been treated as an absolute quantity. Matters change, however, when velocities approach that of light, but these were physically encountered only when the study of elementary particles began.

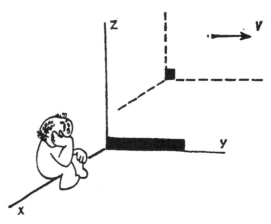

That is about all that had to be said about the concept of length in relativity theory. Yet the relativistic approach to the problem is so unusual that it is worth giving special attention to a question which is frequently asked: Does a body actually grow shorter or is the Lorentz contraction merely a visual effect?

---

that the length of a body moving through the ether contracts. The physical meaning of the contraction formula, as of the whole of Lorentz's hypothesis, however, is entirely different from Einstein's theory. According to Lorentz's theory, for example, we can speak of the absolute length of a body at rest relative to the ether.

Another name you may have come across is the Fitzgerald contraction, which was initially advanced in an attempt to explain the negative result of Michelson's experiment.

The very question shows a lack of understanding of the essence of the matter. It is
correct to say that the Lorentz contraction is
an objective reality. The danger is of a wrong
impression being gained that there exists some

privileged reference frame in which all bodies
have a maximum "proper" length, as against a
contracted length in all other frames.‡ This, of
course, is not the case.

The Lorentz contraction is due solely to the
fact that length is a relative quantity depending
on the reference frame in which it is measured.
To ask whether the Lorentz contraction is
"real" or not is the same as to ask whether the
measured bar is really moving or not.

The latter question does not puzzle us
because we are used to the notion of relative

‡It was this idea that Lorentz postulated in his hypothesis
when he assumed that motion relative to the stationary
ether caused a contraction of length.

velocity, but the relative nature of length is difficult to visualise, and that is what frightens the uninitiated.

Some people claim that acceptance of the concept of the relativity of length contradicts philosophical materialism. Such declarations are due to ignorance both in physics and philosophy. They would not be worth special attention if they were not a reflection of the usual reluctance of people to alter customary notions. Unfortunately, the world is so made that it requires a certain amount of brain work to comprehend it. The latter philosophical remark refers even more to the concept of time.

The result of the accurate analysis reads: *The time lapse between the two events is the least in the reference frame in which the events are seen to take place at one point.* The meaning of this statement may seem rather obscure, but a simple example will clarify it.

The most difficult question. The relativity of time.

Two successive flashes of light occur in a carriage of the Moscow–Leningrad express. According to a clock in the train, the time-lapse

Δt between the flashes is, say, 10 hours. In the "train" frame of reference the flashes occurred at the same point and the "train" clock, naturally, measured the time according to the train frame. If, now, we wish to record the time of the two flashes in a frame of reference attached to the ground according to clocks located at the precise points where the flashes occurred, we will need two clocks, as the flashes occurred in different places. If at the time of the first flash the train clock showed the same time as clock A on the platform of the Moscow railway station, at the moment of the second flash the

train clock will be found to be slow as compared with clock B on the platform of the Leningrad terminal, which is synchronised with clock A.§

---

§Two clocks at different points and fixed in a given frame of reference are said to be synchronised if they simultaneously show the same time. The simultaneity, of course, is determined relative to the same reference frame. From the point of view of an observer in another frame the two clocks may not be synchronised.

In other words, if we compare the movement of a travelling clock with the movement of several stationary synchronised clocks we find that it is slow with respect to the latter.

It should be noted that the reference frames attached to the train and the ground were in

essentially different conditions, for one clock in the train was compared with two clocks on the ground. If we change the experiment and imagine a very long train with a row of synchronised clocks[1] in it and a platform with one clock, we will find that the platform clock is slow compared with the train clocks. That is why it is wrong to say that time passes slower in a moving reference frame. In fact, this contradicts the principle of relativity. Every inertial frame is as good as the other, which means that it is wrong to say that time passes

---

[1] In this case, of course, the concept of simultaneity, required to define synchronised clocks, as defined in the reference frame attached to the train.

faster or slower in one frame than in another. When we speak of the Lorentz "dilation" of time we mean the assertion given earlier and nothing else.[**]

That the time in one inertial frame is as good as in any other can be illustrated by the following example. Imagine two rockets with radio transmitters on board. Let the pilots be equipped with physically identical clocks, suppose the rockets are flying in opposite directions with a constant relative velocity v, and their transmitters emit a radio signal *every second* according to the *respective pilots' clocks*. The pilot of rocket No. 2, recording by *his* clock the intervals between the reception of the signals from rocket No. 1, will find them to be slightly greater than one second by

$$\Delta t = \sqrt{\frac{c + v}{c - v}}.$$

This "dilation" of the time between the arrival of consecutive signals is explained by the Doppler effect.[††]

[**]In view of the extreme importance of this postulate it is worth repeating it: The time-lapse between two intervals is least in the reference frame in which they are seen to have taken place at one point in space. This is the "proper time", denoted by the symbol $\Delta\tau$. The time-lapse between the two events in any other inertial frame is given in terms of $\Delta\tau$ by the equation

$$\Delta t = \frac{\Delta\tau}{\sqrt{1 - \dfrac{v^2}{c^2}}}.$$

[††]This is an opportune moment to recall some aspects of relativistic theory of the Doppler effect for electromagnetic waves. At first glance it seems very like the classical exposition and there is no basis for "surprising" conclusions.

If pilot No. 2 carried out a rather simple calculation, he would find that, according to his clock, the nth signal had been dispatched at time

$$t_n^{\text{No.2}} = \frac{n}{\sqrt{1 - \dfrac{v^2}{c^2}}}$$

seconds after the first. (Without going into the calculations, we shall take this result for granted.)

But by the clock in rocket No. 1 the nth signal was sent at time $t_n^{\text{No.1}} = n$ seconds after

---

Once again, if a source and a receiver are approaching each other, the frequency entering the receiver is greater than if the latter were at rest. And, as in the case now considered, if the source and receiver are receding, the received frequency is less. All this is very much like the conclusions of the classical theory.

But there is one important difference. It is obvious that, once the ether is discarded and the relativity principle is accepted as valid for electromagnetic phenomena, it is useless to distinguish between the two cases: (i) source moves towards stationary receiver, and (ii) receiver moves towards stationary source. The moment the "absolute reference" frame is junked the distinction loses all meaning and the frequency change depends only on the relative velocity of the source and the receiver.

The difference between the formula for the received frequency as compared with the classical is not so important. More important is that the theory of the Doppler effect is closely linked with one of the most remarkable conclusions of Einstein, that of the slower rhythm of moving clocks. That is why, as mentioned before, Einstein regarded experimental verification of his formula for the Doppler effect as extremely important for the justification of the whole theory. Experiments brilliantly confirmed Einstein's conclusions, although the experimenters themselves neither understood nor accepted the theory.

the first, and pilot No. 2 will declare that
No. 1's clock is slow, for

$$\frac{n}{\sqrt{1 - \dfrac{v^2}{c^2}}} > n.$$

The statement of the problem, however, is
completely reciprocal, since rocket No. 2 is *as
good as* rocket No. 1. Therefore we can obviously
freely interchange the numbers of the rockets,
and pilot No. 1 can vouch with equal right that,
as far as he is concerned, it is the second man's
clock that is slow.

Who of the two is right?

The answer is, both.

In order to make sense of this declaration
we must make sure what pilot No. 1 means
by timing the nth signal from rocket No. 2
according to *his* clock. This time actually
means the reading of a clock synchronised with
the first pilot's clock and located *at the point
where the second* rocket was when it emitted
the nth signal. As compared with this clock,
the second pilot's clock will be slow. Similarly,
when the second pilot declares that the clock
in the first rocket is slow he mentally places a
clock synchronised with his own at the point
where rocket No. 1 was at the time it sent off
the nth signal.

We have arrived at our old conclusion: the
slow clock is the one which is compared with
several synchronised clocks in another inertial
frame.

As it stands, this statement sounds somewhat formal, but essentially it coincides with
the basic definition of the time-lapse between
two events, which is the least in the reference

frame in which the events are coincident in space.‡‡

Frankly, "time dilation" is harder to visualise than length contraction. This is probably due, in part at least, to the fact that the concept of time as such is more difficult to visualise, and partially to the "irreversibility" of the effect.

$$\ell = \ell_o \sqrt{1 - \frac{v^2}{c^2}}$$

What is meant by "irreversibility" can best be explained by the example of length. Suppose we accelerate a rod almost to the speed of light relative to some inertial frame and then bring it to a halt. Taking the formulas of the special theory of relativity for all velocities, the length of the rod will be seen by an observer at rest

---

‡‡The mathematical derivation of the Lorentz time dilation is as simple as that of the length contraction. Consider two reference frames, O and O', with the relative velocity parallel to the x axis. For the frame in which the flashes took place at the same point the square of the spacetime interval between them is $c^2\Delta\tau^2$, as Δx — the distance between the points — is zero. For the frame in which the flashes were observed at different points the square of the spacetime interval is $c^2\Delta t^2 - \Delta x^2$.

As the spacetime interval does not change in passing from one frame to another,

$$c^2\Delta\tau^2 = c^2\Delta t^2 - \Delta x^2,$$

or

$$\Delta\tau = \Delta t\sqrt{1 - \frac{\Delta x^2}{c^2\Delta t^2}}.$$

But as $\dfrac{\Delta x}{\Delta t} = v$ (the relative velocity of the frames), then

$$\Delta\tau = \Delta t\sqrt{1 - \frac{v^2}{c^2}}.$$

in our frame to change according to the graph shown here. At the initial instant the length is l_0. It contracts when the motion begins, till it reaches a minimum when the velocity v is maximum, remaining constant as long as the velocity is constant. Then, when retardation begins, the length increases till it reaches its initial value of l_0 at rest. Nothing is left to indicate that the rod had ever contracted.

With time matters are not so simple. If a clock C is accelerated in an imaginary rocket

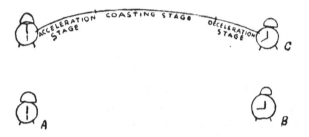

which travels for some time at a uniform velocity v and is then retarded and stopped, it will no longer show the same time as a clock B at its destination, which is synchronised with clock A at the departing station. Clock C will be found to lag. The rate of the clock is the only reversible quantity, for after the flight clock C will run just as it did before (synchronous with clocks A and B). The duration of the journey as measured by it, however, will be less than according to clocks A and B. Here, too, we assume that if the acceleration was not very great, the change in the rate at any given moment is given to a fair degree of accuracy by the formula of the special theory of relativity

$$\Delta \tau = \Delta t \sqrt{1 - \frac{v^2}{c^2}}.$$

Actually, though, neither the length of a body moving with acceleration nor the rate of a clock carried on such a body can be determined by means of the special theory of relativity.

The special theory deals only with inertial frames and, strictly speaking, in all our reasoning the conclusions of the special theory were arbitrarily extended to more general cases. It is generally accepted, however, that if accelerations are, in a certain sense, small,[§§] this can be done. To be sure, some scientists object to this, but we shall blindly follow the majority.

Once again, strictly speaking the problem we have just discussed does not belong to the realm of the special theory. A complete solution is possible only in the framework of the general theory of relativity.

And another very important remark. We took it for granted that an observer in an inertial frame of reference comparing the rate of his clock with that of a clock in an accelerated frame could, with a fair degree of accuracy, make use of the formula given here, which belongs to the special theory of relativity.

[§§]The meaning of this sentence may seem rather obscure, as we are unable to go into a detailed analysis of what is meant by "small" or "large" accelerations. A precise mathematical criterion of the smallness of an acceleration is necessary. In the absence of such a criterion we have discreetly employed the words "in a certain sense".

Let us now take it for granted that in solving a similar problem an observer in a non-inertial frame, whose clock moves with an acceleration, cannot employ the formulas of the special theory.

The "main" paradox of relativity theory: the clock paradox.

And now let us see what is meant by the so-called clock paradox.

Suppose two clocks are carried away from each other with a relative velocity approaching that of light and then brought back together again. From the point of view of observer A,

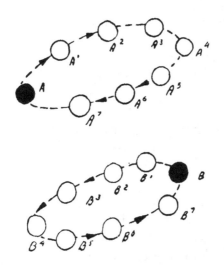

clock B was travelling and its rate was slower. Hence, when they meet, clock B should be slow. Observer B, of course, is fully entitled to reason in the same way, and he declares that it is clock A that should be slow. After the trip clocks A and B are at the same point. The difference in the time they show is an absolute quantity, therefore only one of the two observers can be right.

After our introduction the answer should be obvious. The observer whose clock was accelerated (suppose it is *B*) *has no right* to apply the special relativity theory. If he doesn't know general relativity, then he can say nothing about the rate of clock *A*. Observer *A*, on the other hand, can, as an approximation, employ special relativity, if the acceleration of *B* was not too great. He will conclude (and will be right) that clock *B* is lagging, and he can calculate by how much.

If the acceleration of *B* was "large", however, then the special theory gives nothing for a correct answer. The general theory says that in this case, too, clock *B* will lag.

And finally, if both *B* and *A* moved with acceleration, the whole problem must be relegated to the general theory, and many variants are possible.

Thus, the answer to the apparent paradox lies in the "unequality" of the two clocks, for if they separated and then met, at least one of them had to move with acceleration.¶¶

¶¶Our reasoning gives only a qualitative explanation of the clock paradox. As to the mathematical interpretation of the difference between the two clocks, opinions differ. Here, as in the concluding chapter where we shall mention the clock paradox again, we will stick to the more widespread notion.

CHAPTER XV,

the content of which should atone for its faults

EINSTEIN. THE LAWS OF MECHANICS
(mass and energy)

A person brought up on classical concepts is bewildered by Einstein's mechanical laws, though actually there is much more in common between relativistic and Newtonian mechanics than might seem at first glance.

To begin with, Newton's first law remains unchanged in relativistic mechanics: in an inertial frame of reference the momentum of a body not subjected to external forces does not change.

More talk.

The third law — the equality of action and reaction — is also valid in relativistic mechanics, and we can assert that if two bodies are interacting their total momentum does not change.

Actually, the second law of mechanics also remains the same, and force is characterised by the rate of change of momentum:

Newton's second law in relativistic mechanics.

$$F = \lim_{\Delta t \to 0} \frac{\Delta P}{\Delta t}.$$

But though the content of the second law is the same, its specific form is markedly different. We shall have to take it for granted that in relativistic mechanics the momentum of a body is given by the expression

$$\vec{P} = \frac{m\vec{v}}{\sqrt{1 - \dfrac{v^2}{c^2}}}.$$

It is beyond our scope to investigate how this formula is derived and we shall merely note that the definition of momentum looks fairly natural and plausible. For one, at velocities far below that of light we obtain (as we should) the familiar classical expression for momentum:

$$P = mv.$$

On the other hand, when the velocity of a body approaches that of light its momentum tends to infinity. This, too, is understandable, for it corresponds completely to the notion that no material body can be accelerated to the velocity of light.

The graph here shows how the actual momentum compares with the approximate classical expression. The solid line is the relativistic momentum curve, and the broken line, the classical. Even at what would conventionally be regarded as very high speeds the relativistic formula coincides almost exactly with the classical.

Thus, at 30 km per sec* the momentum as calculated by the classical formula is only a half a millionth of one per cent less than the actual quantity.

It is obvious, therefore, that even in space flight design, relativistic effects are not

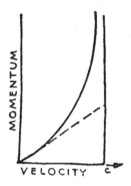

essential. It would, of course, be even more useless to apply the exact formulas of relativity theory in investigating the much slower motions we encounter in everyday engineering problems. For all of these cases Newtonian mechanics gives an excellent approximation.

In our age, however, engineers are encountering more and more purely technical problems to which relativistic mechanics have to be applied. Electrons, protons and other elementary particles are speeded up in modern accelerators to velocities very nearly approaching that of light. An electron can be accelerated by means of the relatively small potential of one million volts to 0.92 c. At this

*This is almost three times more than the escape velocity of 11.2 km/sec needed to overcome the Earth's gravitational pull.

speed the momentum as calculated by the classical formula is only one-third of the actual value. Needless to say, in designing particle accelerators the formulas of relativistic mechanics are employed. Thus, in our time Einstein's theory has found engineering application. Neither is it inopportune to state that experience has confirmed Einstein's formulas.

But get back to Newton's second law as applied to relativistic mechanics. If we say that the momentum of a body is defined as the product of mass of the body times its velocity then, taking another step forward, we find that the mass itself depends on the velocity:

$$M = \frac{m}{\sqrt{1 - \dfrac{v^2}{c^2}}},$$

where m is the mass of the body at rest, called the *rest mass*.

Rest mass. Remarks concerning the concept of mass.

From the point of view of classical physics, this relation is incredible, yet there is nothing absurd in it.

The world is so established that velocity affects mass. That this is not perceptible at small velocities (Newtonian mechanics) is no reason to reject the possibility offhand.

True, one might challenge our definition of mass and declare, for example, that momentum should not be taken as the product of mass and velocity, that the "true", "real" mass of a body is its rest mass m. Still, the definition of mass given above is probably the most logical. No one, of course, is barred from offering other definitions, but physicists will hardly consider them.

Mass characterises the inertia of a body, i.e., its tendency to retain its initial state of rest or motion (in an inertial frame of reference). Inertia, it was found, depends on velocity, and this must be taken into account in the definition of mass. But maybe there is no sense in introducing the concept of mass at all?

In this connection it would be worth remembering that physical concepts are not ordained from above, they are not something established once and for all and existing by themselves. Physicists introduce concepts in order to provide a better and more logical description of the real world. New discoveries may bring to light unknown properties which require a new approach to physical concepts.

This was the case with mass. Thus, in classical physics mass could be defined as a coefficient of proportionality between force and acceleration. This proved wrong and applicable only to small velocities, when Newtonian mechanics is valid and mass, for all practical purposes, does not depend on velocity.

In relativistic mechanics, as a rule, acceleration does not coincide in direction with the vector of the applied force. Only in two cases do they coincide: when the force vector is either in the direction of, or perpendicular to, the velocity vector. In all other cases acceleration is *not parallel* to force.

More, even when the acceleration and force vectors are parallel, it is wrong to speak of some uniform coefficient of proportionality between force and acceleration. Actually, it is easier to deflect a body from its path (acceleration is perpendicular to velocity) than to increase the velocity (acceleration is parallel to velocity).

Therefore, mass should be defined in terms of momentum, applying the most general form of the second law of mechanics. Incidentally, Newton himself formulated the second law in the way it is given at the beginning of this chapter, not in its familiar "school physics" form of ($\overrightarrow{F} = \overrightarrow{ma}$).

On the whole, the new conception of mass may cause surprise, but it should not cause consternation. Einstein's mechanics, though, has another surprise in store.

Einstein developed the following formula for the kinetic energy of a body:

$$E_k(v) = \frac{mc^2}{\sqrt{1 - \dfrac{v^2}{c^2}}} - E_0,$$

The major conclusion of the special theory of relativity: the relation between mass and energy.

where E_0 is a constant that can easily be determined. By definition, the kinetic energy $E_k(v)$ obtained by a body is equal to the work done by the external forces to accelerate the body from rest to the velocity v. It follows, therefore, that the kinetic energy of a body at rest is zero. A glance at the formula shows that this condition is fulfilled when

$$E_0 = mc^2.$$

The relativistic expression for kinetic energy now takes the form

$$E_k(v) = mc^2 \left(\frac{1}{\sqrt{1 - \dfrac{v^2}{c^2}}} - 1 \right).^{\dagger}$$

---

†It should be pointed out that often the quantity

$$E = \frac{mc^2}{\sqrt{1 - \dfrac{v^2}{c^2}}}$$

is defined as the relativistic kinetic energy.

At small velocities, when $\dfrac{v}{c} \ll 1$, the relativistic expression of kinetic energy turns into the classical formula $E_k = \dfrac{mv^2}{2}$, which is easily proved.[‡]

On the other hand, at velocities approaching c, the kinetic energy tends to infinity, as it should. This is all very well, but still the formula arouses some suspicion, and here is why. The work done by external forces on a body must equal the difference between its total energy in the initial and the final states. If all the work is used up on imparting kinetic energy to a body then, naturally, the kinetic energy determines the total energy increase:

$$E_k(v) = E_{total}(v) - E_{total}(0).$$

In classical mechanics the total energy of a stationary body is usually immaterial. In solving problems, only those forms of total energy have to be taken into account which

In the final analysis this is a question of terminology. Without going into a further discussion, we shall continue to abide by the definition given here, as it explains the relation between mass and energy in simpler form, though maybe not so logically.

[‡]At $\dfrac{v}{c} \ll 1$,

$$E_k = mc^2 \left(\frac{1}{\sqrt{1 - \dfrac{v^2}{c^2}}} - 1 \right) \approx mc^2 \left(\frac{1}{1 - \dfrac{v^2}{2c^2}} - 1 \right)$$

$$\approx mc^2 \left(1 + \frac{v^2}{2c^2} - 1 \right) = \frac{mv^2}{2}.$$

It is worth taking advantage of this example to stress the purely circumstantial character of the expressions "small" and "large". In our case a velocity is small if $\dfrac{v}{c} \ll 1$; thus, 100 km/sec is a small velocity.

change during the motion (potential energy, for instance). In any specific mechanical problem the energy of a body can be defined such that the energy of the free body at rest is zero.

In relativistic mechanics, however, the kinetic energy of a body is expressed as a difference of two terms, and the "point of origin" for the energy count is, for some reason, not zero. At this stage we may, on a purely formal basis, ascribe to any stationary free body an energy $E_0 = mc^2$. Then,

$$E_k(v) = E(v) - E_0 = (M - m)c^2,$$

where

$$M = \frac{m}{\sqrt{1 - \dfrac{v^2}{c^2}}}$$

is the variable mass of the body. The total energy of a body at rest can thus be expressed by the formula

$$E = Mc^2.$$

What is this, a mathematical freak? A purely formal circumstance? Does mc^2 have any physical meaning or is it merely "energy" in quotes?

The logics of Einstein's theory led him to the conclusion that the energy of rest $E = mc^2$ was a real physical quantity. Everybody actually carries within itself such an energy. We should point out here that the smooth words about "the logic of his theory" should not hide from us the daring of Einstein's thinking, something which defies description.

Einstein himself regarded this as an important corollary of his theory.

"The mass of a body is a measure of its

energy," he wrote in 1905. "If its energy changes by ΔE, then its mass changes by $\Delta E/(9 \times 10^{20})$, if energy is measured in ergs and mass in grams. Is it not possible that one could test this hypothesis for bodies with an energy content changeable in a high degree (for example, the radium salts)?"

Thus, to every mass there corresponds some kind of energy, and to every type of energy there corresponds some mass. The relation between them is given by the formula $E = mc^2$. The mass of a heated body is greater than the mass of the cold body. Reciprocally, in cooling or giving off energy in some other way a body loses mass. Every process involving the release of energy is accompanied by a loss of mass and, reciprocally, any body or frame of bodies absorbing energy increases in mass. In short, any giving off or absorption of energy is accompanied by a change in mass.

Thus, for example, the rest mass of two hydrogen atoms must be greater than the rest mass of a twin-atom molecule of hydrogen since atoms joining into molecules give off energy, which means a loss of mass:

$$H + H = H_2 + Q; \quad 2m_H > m_{H_2}.$$

In any exothermic chemical reaction, in which energy is given off, the mass of the reaction products is less than the mass of the reactants. Conversely, in an endothermic reaction, in which energy is consumed, the mass of the reaction products is greater than that of the reacting substances. The simplest example of an endothermic reaction is the dissociation of a hydrogen molecule into atoms:

$$H_2 + Q = H + H; \quad m_{H_2} < 2m_H.$$

Of course, no one ever takes account of the change in mass in the formation of a hydrogen molecule. The most accurate measurements would fail to reveal any such change in ordinary chemical reactions, and the law of conservation of mass holds good for all chemical reactions.

Also, no change in mass will be detected if we weigh a cold lump of iron and then weigh it again after heating, although the change in energy is readily perceptible.

Why is it that we fail to observe any sensible

change in mass in chemical or other processes involving substantial energy changes?

The answer is apparent from an investigation of the formula $E = mc^2$. A very small change in the first multiplier (mass) will cause a very great change in E.

Mass is much more "expensive" than energy. One gram of mass is equivalent to 9×10^{20} ergs of energy. Conversely, of course, one erg of energy is equal to the infinitesimal mass of $1/(9 \times 10^{20})$ gram. The energy corresponding to one gram of mass is equivalent to the kinetic energy of a rocket weighing 1,500 tons accelerated sufficiently to escape the Earth's gravitation (11.2 km/sec).

Sportsmen claim that a football player may lose as much as ten pounds in the course of a game. This may be so, but hardly any centre forward realises how much energy he loses with this mass. Transferred into energy, it would be sufficient to kick a football weighing five million tons clean beyond the Earth's gravitational field.

As to the energy evolved (or absorbed) in chemical reactions, the change in mass involved is so infinitesimal as to be totally beyond the range of the most delicate instruments, even if their sensitivity were increased a thousandfold. The same goes for any increase in the mass of a heated body, which is given by a digit lying so far to the right of the decimal point as to belong rather to the realm of speculative curios.

Nuclear energy. The mass defect.

The situation, however, changes markedly when nuclear reactions are involved. Back in 1905 Einstein considered radioactive processes as a means of verifying mass change. Then it was a pure hypothesis. Today theory has been confirmed by the results of the many nuclear reactions known to us. The energy liberated or consumed in nuclear reactions is hundreds of thousands and millions of times greater than the energy output of ordinary chemical reactions, and the corresponding mass changes are also millions of times greater. In the reaction of formation of water, for example, 136,000 small calories are liberated to every two moles of hydrogen and one mole of oxygen (i.e., 36 grams of matter):

$$2H_2 + O_2 = 2H_2O + 136,000 \text{ cal.}$$

In the nuclear reaction of the formation of helium from lithium and hydrogen

$$Li^7 + H^1 = 2He^4 + Q,$$

to every 7 grams of lithium nuclei and 1 gram of hydrogen nuclei some 5,000,000,000 calories are liberated. With such an energy output mass change is readily observable.[§]

Even in nuclear reactions, however, the change in mass is usually within fractions of one percent. Like a thrifty owner, nature is very sparing with its energy stocks and even in such cataclysmic phenomena as nuclear explosions it expends a very small portion of its stocks. To illustrate this, here is the energy balance of the lithium-hydrogen reaction.[¶]

$$Li^7 + H^1 = 2He^4 + Q.$$

Accurate measurements give the following figures for the masses of the respective atoms:

lithium (Li^7) = $7.01818 \times 1.66 \times 10^{-24}$ g,

hydrogen (H^1) = $1.00813 \times 1.66 \times 10^{-24}$ g,

helium (He^4) = $4.00389 \times 1.66 \times 10^{-24}$ g.

Let us compute the masses of the reacting and the reaction products:

$$Li^7 + H^1 \rightarrow 2He^4,$$

[§]Incidentally, we may sometimes read that the evolution (or absorption) of huge quantities of energy in nuclear reactions *is caused* (or explained) by the change in the mass of the substances entering the reaction. This statement is wholly incorrect. The apparent change in mass (the mass defect) *indicates* that the reaction among nuclei takes place with the evolution or absorption of a tremendous amount of energy, but does nothing to explain *why* so much energy is released in nuclear reactions. The latter question can be answered only when the nature of nuclear forces is investigated.

[¶]The multiplier 1.66×10^{-24} is the mass of a proton.

$$7.01818 \times 1.66 \times 10^{-24} + 1.00813 \times 1.66 \times 10^{-24}$$

$$\rightarrow 2 \times 4.0039 \times 1.66 \times 10^{-24} \text{ g,}$$

whence

$$8.02631 \times 1.66 \times 10^{-24} \text{ g}$$

$$\rightarrow 8.00778 \times 1.66 \times 10^{-24} \text{ g.}$$

In the left term we have an excess of mass equal to 3.08×10^{-26} g. The energy liberated in the course of the reaction (which is added to the right-hand member of the equation) must correspond to this mass, whence

$$Q = \Delta m c^2 = 3.08 \times 10^{-26} \text{g} \times 9 \times 10^{20} \frac{\text{cm}^2}{\text{sec}^2}$$

$$= 2.72 \times 10^{-5} \text{erg.}$$

The energy evolved in the reaction constitutes the kinetic energy of the newly formed helium nuclei (α-particles).

Experimental data confirm completely the theoretical calculations of this and hundreds of other nuclear reactions. By accurately determining the masses of all the atomic nuclei it is possible to forecast the course of any nuclear reaction, stating whether energy will be released or consumed, and precisely how much.

In our example we predicted the liberation of energy, and the reaction could, in fact, be used as an extremely powerful energy source: two reacting nuclei of lithium and hydrogen release a tremendous amount of energy, 2.76×10^{-5} erg! Remember that we are speaking of only two atoms. In ordinary chemical reactions the corresponding energy release is less by a factor of several millions and constitutes about 10^{-11} or 10^{-12} erg. To trigger a nuclear reaction the nuclear energy barrier has to be overcome, and for this energy must be expended. True, the net gain more than compensates for the expenses, but the barrier exists and it has to be surmounted. The graph here represents a schematic energy diagram for nuclear reactions.

The liberated energy is usually considerably greater than the activation energy ($\Delta E \gg E$). The scale of the diagram does not convey a true idea of this difference, for in reality ΔE may be dozens of times greater than the activation energy.

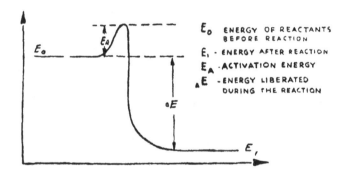

E_0 ENERGY OF REACTANTS BEFORE REACTION

E_1 - ENERGY AFTER REACTION

E_A - ACTIVATION ENERGY

ΔE - ENERGY LIBERATED DURING THE REACTION

If conditions are created in which part of the released energy is used to surmount the energy barriers of the atoms that have not yet entered into the reaction, we have a chain reaction. Without going into a detailed discussion of nuclear reactions, let us restrict ourselves to an explanation of the statement that, for the overwhelming majority of nuclear reactions, a high activation barrier is natural and nothing else could be expected.

If such a barrier had not existed, all elements would long since have interacted and "fallen into energy wells". Nuclear reactions would have been impossible in the absence of the necessary "raw material".

On the other hand, in conditions of continuous nuclear reactions it would obviously have been impossible for such highly organised matter as living and intelligent creatures to appear. We should be happy, therefore, that on Earth nuclear reactions take place as a rule only in artificial conditions. There is one exception however: the spontaneous fission of natural radioactive elements, the discovery of which marked the beginning of the nuclear age.

Inside stars, however, nuclear reactions are taking place continuously and violently. A few figures will offer an idea of the incredible energy produced in stellar thermonuclear "piles". Our Sun is a very middling star, yet the energy radiated by it every second is equivalent to some 5,000,000 tons of mass.** The Earth's share is equivalent to about two kilograms of mass per second, thanks to which it gains in weight some 170 tons a day.††

Fragmentary as they are, the examples cited here offer an idea of the importance of the theory of relativity in our time.

**This should not cause alarm: at this rate it would take the Sun some 6,000,000,000,000 years to lose half its mass.

††The actual gain, though, is considerably less, as the Earth also radiates energy into space.

CONCLUSION,

in which the author bids the reader farewell

Realisation of the fact that the reader has reached thus far in this book is a source of pleasure to the author. He furthermore hopes that the reader shares this feeling, though for another reason. In any case, there is probably no use in wasting too much time on the finishing touches.

It is hardly necessary to expound any further on the importance of the theory of relativity. It should be apparent by now that practically the whole of modern physics is connected, to a greater or lesser extent, with relativity theory; nuclear physics and astrophysics are unthinkable without it.

Such declarative statements are rarely convincing. Moreover, if after all that has been said the reader has failed to see the import of Einstein's theory in contemporary physics, it is hopeless to expect that two or three more pages will save the situation. If, on the other hand, this has been shown, then all the less need for any further exposition.

As to the experimental proofs of the theory, it would undoubtedly have been useful to consider several examples and see how relativity theory explains, say, the aberration of light. Unfortunately, an adequate analysis of these experiments is impossible without going deep into mathematics; it is hardly worth increasing the number of illustrations. But one general statement ought to be made.

In the 55 years that have passed since Einstein's first paper appeared, not a single experimental fact has been found to contradict the theory. On the contrary, the whole complex of available experimental data is beautifully explained by relativity theory. Moreover, and significantly so, Einstein's theory has predicted many new and unexpected effects which were later observed. The most vivid example of such prediction is the equivalence of mass and energy.

Before putting the final period, I should like to say that even apart from its purely scientific value, Einstein's work, in which extraordinary physical ideas are clothed in the flawless, exquisite dress of mathematics, is characterised by the inner logic and beauty which distinguishes outstanding works of art.

Our discourse is over. The following chapter was included only as a concession to the vogue. Its conclusions, however, can be regarded as an amusing example of some of the laws of relativistic mechanics, which also justifies its appearance in the book.

CHAPTER XVI,

the last, and in some respects a
heretical one. It anathematises
photon rockets and sets forth
the author's ideas on day-
dreaming. After this the over-
patient reader will probably
fling the book away with a sigh
of relief

PHOTON DREAMS

*Manilov watched the chaise disappearing in
the distance.... Then his thoughts passed imper-
ceptibly to other subjects, and goodness knows
where they landed at last. He mused on the bliss
of life spent in friendship, thought how nice it
would be to live with a friend on some bank of
a river, then a bridge began to rise across the
river, then an immense house with such a high
belvedere that one could see even Moscow from it,
and then he dreamed of drinking tea there in the
evening in the open air and discussing agreeable
subjects. Then he saw himself and Chichikov
driving up in a smart turn-out to attend some
party, where they charmed everyone with their
polite manners... and on and on till goodness
knows wither his dreams carried him.*

GOGOL

I am very sorry, but I simply must write about
photon rockets. In our time it seems awkward
not to mention them in a book on relativity
theory. Perhaps nothing has captivated the

An introduction to ex-
plain what will follow
and several general
remarks.

imagination of our generation more than the idea of photon rockets capable, it is claimed, of reaching the most distant stars of the universe.

The idea, in fact, competes successfully with stories about the future of cybernetics or the prospects of harnessing thermonuclear energy. The problem is neither more nor less than that of space vehicles capable of travelling at speeds, relative to the launching pad, approaching the velocity of light.

It is inevitable, therefore, that relativity theory is called upon, for the motion of these hypothetical spaceships will be governed by the laws of relativistic mechanics. People discussing photon rockets always appeal to Einstein's theory, just as very ordinary persons like to mention some celebrated acquaintances of theirs. But before attempting to judge the problem on its merits let us try and solve a sort of psychological puzzle.

What has caused the tremendous, sensational success of the idea of photon rockets as vehicles for space conquest? Success demonstrated by innumerable science-fiction works whose authors with characteristic lack of imagination, as Lord Kelvin caustically remarked very long ago, so easily fill the Milky Way with sundry stellar ships.

There are probably two main reasons for this. First and foremost is the fascination of the idea of conquering the boundless expanses of the universe. Secondly, the idea is doubly attractive in our time, when the Earth's gravity pull has been overcome, manned spaceships have circled the globe, an attempt to land on the Moon will undoubtedly be undertaken in the near future, and astronautics is rapidly becoming an applied science. With all this in mind, it is, naturally, difficult to swallow the idea that no technological achievements will ever make it possible for man to reach the distant stars.

Alackaday, a dream, however wonderful, is but a dream, and so far we have no grounds for speculation that mankind will ever be able to build rockets capable of commuting between stellar worlds, to say nothing of galaxies. I myself regret this conclusion and would be happy to learn of any specific propositions which would make flight to the distant stars feasible. Unfortunately, speculations to date concerning photon rockets constitute nothing but wishful thinking.

I shall attempt to lay forth all the pros and cons with complete objectivity so that the reader could judge for himself whether this uncomfortable conclusion is true.

What is a "photon rocket" anyhow? People in the know claim that photon rockets will be accelerated by the reaction force developed by a powerful stream of electromagnetic radiation quanta, that is, photons. Directed electromagnetic radiation carries momentum; as the total momentum of the closed system "rocket + radiation" must be zero, the rocket is imparted an equal and oppositely directed momentum.

There is nothing new in this as compared with the well-known principles of rocket motion. The unusual feature is photon reaction propulsion. Such an ambitious motor is chosen for stellar flight because the most efficient fuel is that which develops a jet with the highest possible velocity relative to the ship, that of

The reader is informed in advance of the conclusion drawn in this chapter.

light being the cherished goal.* Attainment of the velocity of light assumes the ejection of mass in the form of electromagnetic radiation quanta — photons — as the velocity of light can be achieved only if the rest mass of the accelerated particle is zero.†

Incidentally, we could discuss electron, proton or meson rockets with equal success. If, for example, the speed of the escaping electrons relative to such a rocket is approaching that of light, the loss in momentum, per unit of ejected mass, as compared with a photon rocket, will not be so great. As far as imagining rockets with near-light velocity are concerned, acceleration by electrons is even better to envisage, though, of course, the choice of a subject for day-dreaming is a matter of taste.

First let us see why velocities approaching that of light inevitably accompany dreams of stellar flight.

The closest star, Proxima Centauri, is 4.2 light-years away. Accordingly, the time needed

---

*The speed of light can be attained only if the whole mass of the propellant is transformed into electromagnetic field quanta ejected by the motor. If not all the fuel burns in the motor and some of the combustion products remain in the form of a ballast, which has to be discarded without adding to the momentum, more efficient performance could be obtained by ejecting the whole mass of the propellant through the rocket motor, the speed, necessarily, being much less than that of light. All these, however, are subtleties which do not concern us.

†We could even amuse ourselves with the ideas of neutrino rockets. The rest mass of a neutrino is zero, hence neutrino propulsion is as good as photon propulsion. The idea of a neutrino rocket, I am afraid however, lies "on the other side of good and evil".

to reach it travelling with a velocity v is

$$t = 4.2 \times \frac{c}{v} \text{ years.}$$

Thus, a trip even to our closest neighbour in the universe entails speeds relative to the solar system comparable with the velocity of light, otherwise it may last tens of thousands of years (some 12,600 years at the suitable "inter-planetary" speed of 100 kilometres per second). The inconvenience of such long journeys need hardly be explained, hence the need for rockets flying with more or less the speed of light.

Let us see what we need to embark on such a journey, limiting ourselves for the time being to our closest stellar neighbours. If we could build a ship capable of accelerating to 100,000 km/sec, the whole flight would take some thirty years or so. This is quite a span, but, on the whole, acceptable, so we will be content for the time being with such a "slow" rocket.[‡] First we have to consider our useful mass, that is, the total

Sub-relativistic rockets and trips to the closest stars are also sheer fantasy, but at least within reason.

[‡]This is also convenient because, at 100,000 km/sec, the good old Newtonian mechanics can be used to describe the rocket's motion with a fair approximation. At this speed the mass of the rocket is six percent higher than the rest mass:

$$M = \frac{m}{\sqrt{1 - \left(\frac{1}{3}\right)^2}} = \frac{m}{\sqrt{0.89}} \approx 1.06\, m.$$

Accordingly, the rocket's momentum is also six percent more than that given by Newton's formula. Also we find that the rocket's kinetic energy

$$E_\kappa = mc^2 \left(\frac{1}{\sqrt{0.89}} - 1\right) = 0.06\, mc^2$$

mass of the rocket less the fuel. There is plenty of room for conjecture, but 100,000 tons seems to be about the smallest more or less plausible figure (we should try to be reasonable even in our fancies).

One hundred thousand tons! This seems a tremendous figure. But then, big ocean liners have displacements of fifty, eighty and more thousands of tons. It is hardly feasible to think of a rocket smaller than an average ocean liner, if only because, as we shall soon see, it needs an enormous amount of fuel, which has to be stored somewhere. And what about the hull? It will have to be tougher than that of a battleship. The worst naval battles will be as sling-shot fights between children in comparison with the ceaseless bombardment to which the rocket will be subjected in flight.

Furthermore, the rocket has to be packed with scientific apparatus and flight control mechanisms, to say nothing of the motor. Undoubtedly nuclear-powered, the motor will have to be surrounded by a protective shield weighing (even in our dreams) several thousand tons. In short, the boldest enthusiasts must concede that in assuming a useful mass of 100,000 (10^5) tons, we have been over-optimistic. But even so, the utter futility of the project will shortly be apparent.

---

exceeds the energy calculated according to the classical formula by some 8 percent, since

$$E_{classical} = \frac{mv^2}{2} = \frac{mc^2}{18}.$$

The disparities are not so great and we can calculate the motion of such a rocket without involving relativity theory.

As a further concession to our more zealous fantasts, let us imagine that the hull of our ship can withstand erosion by cosmic dust and afford protection from cosmic radiation. Of course, no super-hull could shield the crew from those hazards, but suppose we have found a way out.

The thing is that, as soon as we begin tackling the motor problem, everything else recedes to the background (though probably with equal justification we could say "if we start tackling the protection problem we will have our hands full without the motor").

Problem No. 1 is that of fuel. Any chemical fuel must be rejected offhand. At 100,000 km/sec a kilogram of mass has a kinetic energy of 5.4×10^{14} kilogram-metres. This energy has to be "paid for". Even assuming a motor whose efficiency is unity and neglecting the action of external forces, to accelerate every kilogram of the rocket's mass we would have to burn enough fuel to liberate 5.4×10^{22} ergs.[§]

It is difficult even to appreciate the vastness of this quantity by terrestrial standards. The volume of conventional fuels needed to produce that amount of energy would run into tens, hundreds, and thousands of cubic kilometres. The only feasible energy source, therefore, is nuclear energy.

At first glance this seems to save the day, for only 60 grams of matter would have to be transformed into electromagnetic quanta for the acceleration of every kilogram of mass.

Processes in which all of the interacting

For a very important reason we do away with interstellar matter.

[§]Actually much more is needed, for the unburned fuel also has to be accelerated together with the payload. But that is neither here nor there.

matter changes into radiation are known. These are the annihilation reactions of elementary particles with their respective anti-particles. Thus, in the "electron-positron" reaction the two particles "burn up" completely with the formation of two gamma-quanta.

Comforting speculations concerning "antifuel".

However, even the most fervent imagination must concede that the hopes of using such reactions in engineering are nil, if only because it is impossible to imagine a reservoir for antimatter. The anti-particles would instantaneously react with the walls, and the rocket and crew would instantaneously be transferred to "extra-stellar worlds". But maybe it would be possible to store a sufficient amount of antimatter in a so-called magnetic bottle, produced by an extremely powerful magnetic field? In this case the fuel would be kept away from the walls. Well, no one can be prevented from imagining what he will. In the middle ages people were equally justified in supposing that a genie could be sealed in an ordinary (non-magnetic) bottle.

However, as long as our destination is only Proxima, we may, reluctantly, accept conventional nuclear fuel. We can count on the already harnessed reactions of fission of heavy nuclei or on the prospective reactions of fusion of light nuclei. Considering feasible nuclear propellants, several kilograms of nuclear fuel, say 10 kg, would be enough to accelerate one kilogram of mass to 100,000 km/sec.[¶] Remembering that in the course of the journey our ship will have to be accelerated at least twice (on leaving the

---

[¶]Incidentally, calculations show that such a fuel would have to possess a fantastically high energy-output as compared with known nuclear fuels.

Earth and leaving Proxima) and decelerated
as many times (on approaching the respective
destinations) and that the fuel for this must
be carried with us, we find that we will need

at least 10 tons of nuclear fuel per kilogram of
useful mass of the rocket.

Thus, if the useful mass is 10^5 tons, the total
initial mass must be at least 10^9 tons. This
compares with the mass of an average-sized
metallic asteroid about 1/10th cu km in volume.
Let us estimate the energy of the jet necessary
to accelerate the rocket at the rate of 1 m/sec^2.
Assuming the jet to consist of massive particles,
at the reasonable ejection speed of around
100,000 km/sec the kinetic energy developed by

Some hopeful figures.

the jet in one second must be expressed by the hardly comprehensible figure of 10^{27} ergs, A photon motor is hardly a way out. The power of a photon jet providing the necessary thrust is 3×10^{27} erg/sec.

There are no known processes on Earth which would develop that much energy per second. The whole globe receives 550 times less energy from the Sun per second. The necessary power could be generated by the complete "burning" of 1,100 kilograms of mass every second or, in terms of uranium, 1,300 tons of uranium. This is the energy that would be released by the explosion of a million atom bombs.

In all our estimates we took the smallest possible figures, including the acceleration. The result, however, is so staggering that we can allow ourselves the luxury of reducing it by a factor of 100 or 1,000 or, if you wish, 10,000: it is equally impossible to imagine a nuclear-powered rocket motor with a capacity of 10^{27} erg/sec as with a capacity of 10^{23} erg/sec. In either case the released energy would instantaneously reduce the rocket to ashes.

We can dream of materials capable of withstanding enormous temperatures and pressures produced in the motor, but all available physical experience, all our notions concerning the structure of matter drive us to the inevitable conclusions that substances with such fabulous properties are impossible. The requirements imposed on the material of the motor are much greater than on the rocket hull. Yet we are unable to imagine how to protect the ship from collisions with meteoric particles. In short, the whole idea can confidently and finally be passed back for further exploration in the devil-may-care department of science fiction.

Yet even these dreams seem as sober speculation as compared with visions of rockets travelling at nearly the speed of light.

Photon-rocket dreamers are not satisfied with the unassuming sub-relativistic rockets of the type just described. They want stellar ships flying with the speed of light, or nearly so. Why are they so adamant in this demand of theirs? Why not be content with sub-relativistic rockets?

The answer is directly linked, first, with the size of the universe, and second, with the so-called clock paradox.

Our Galaxy measures some 10^5 light-years across. Other galaxies lie hundreds of thousands and millions of light-years away. It would seem, therefore, that even with rockets travelling with the speed of light, no possible or impossible technological achievements can ever take man outside of a tiny island of the universe measuring several score light-years (somewhere around 10^{14} kilometres).

Relativity theory seems to open new horizons (though, generally speaking, the problem of time-flow in an accelerated rocket cannot be solved by means of the special relativity theory alone). It is usually assumed that, with a fair degree of accuracy, for every given instance the connection between rocket time and terrestrial time is given by the Lorentz formula:[**]

$$\Delta\tau = \Delta t\sqrt{1 - \frac{v^2t}{c^2}}.$$

Relativistic rockets with near-light speed. Fantasy beyond the border of reason.

A streamlined discussion of time dilation in a relativistic rocket.

[**]Some scientists (a small minority) object to such an examination and regard it as lawless. The question is thus

This means that we have only to accelerate the rocket to almost the velocity of light to win as much time as we want as compared with terrestrial time. Thus we can solve the problem of reaching distant parts of the universe. True, the gain of time in the acceleration stage will not be great. But, with a "proper acceleration" of the order of the acceleration of gravity, a spaceship can be accelerated within several "proper years" to a velocity $v = \left(1 - \dfrac{10^{-8}}{2}\right)c$, which is sufficient to fly around the whole Milky Way within a reasonable "proper time".[††] From this aspect things seem more or less all right. Though, if the flight range were increased, say, a hundredfold, a lifetime would not be enough even by rocket time standards. But let us keep within the Milky Way.

Naturally, on his return home a space traveller will find that tens of thousands of years have passed and he will have to adjust himself to a new species of humanity. But the loftiness of his task will undoubtedly justify the sacrifices entailed in his flight "into the hereafter".

So all that is necessary is to achieve a

---

open to debate. The author must acknowledge that so far he has been unable to find a work in which the problem of the "dilation of time" in a rocket would be examined quite flawlessly, though this, of course, may be due to his own lack of comprehension and not to any fault of the books.

[††]To an observer on Earth the acceleration time is, of course, tremendous. Thus, given an acceleration $a = 10$ m/sec^2, the time needed to attain a velocity $v = \left(1 - \dfrac{10^{-8}}{2}\right)c$ is 9.6 years of rocket time and 9,600 years terrestrial time.

velocity close to that of light. The chances of this however are about the same as (if not less than) of manufacturing a live man out of a pile of atoms taken from the periodic table. Even supposing people had already built sub-relativistic rockets, mankind would still be infinitely far from producing rockets capable of circumnavigating the Milky Way. This is due to the properties of motion at the upper threshold, the speed of light. For, as velocity approaches that of light, the mass of the rocket will tend to infinity, and the force needed to accelerate it increases accordingly. Hence, the fuel expenditure will rise catastrophically above the "small" figure of sub-relativistic requirements. As to protection against erosion by cosmic dust, at near-light velocities it is an utterly hopeless task.

But suppose the protection problem has been solved. Suppose, further, that we have produced a motor capable of developing the tremendous energy needed to accelerate a vehicle to near-light speed without incinerating itself. Suppose that the thrust is exerted by a beam of light quanta, the most efficient propellant. Suppose, finally, that we have a nuclear superfuel with an efficiency factor equal to unity (all the mass changes into radiation).

Supposing everything possible and impossible.

Supposing that all these fantastic requirements are met, even then to accelerate a rocket to 0.9999 of the speed of light it would take 200 kg of fuel per kilogram of final rest mass of the rocket. Assuming that at least two accelerations and two retardations are necessary, we find that the initial mass of the rocket will have to be 10^9 times greater than the useful mass. This figure speaks for itself.

The speed required to circumnavigate the Milky Way in a reasonable time is

$$v = \left(1 - \frac{10^{-8}}{2}\right)^5 c.$$

If it is to be reached, the ratio of the initial mass to the useful mass becomes 10^{17} to one (!!!). Proceeding from our really small useful mass of 10^5 tons, the total mass will be 10^{22} tons. For your information, the mass of our little planet is 6×10^{21} tons.[‡‡]

It is hardly necessary to cite other difficulties. I think the situation is obvious to any unprejudiced person. The photon enthusiasts (I sincerely regret that I do not belong to that category) will hardly be convinced by all that has been said. Scores of projects could probably be developed for storing anti-matter, protection against cosmic dust and ejecting a photon stream without risk of turning the rocket and its crew into elementary particles at the very beginning of the flight. Yet it seems obvious that in our time, with our level of knowledge, no technologically feasible ideas can be proposed. The projects suggested to date go too far, even as sci-fic studies. The fantast is entitled to operate with the unknown and

---

[‡‡]Lack of space makes it impossible to discuss the idea of a ram-jet type of photon motor designed to reduce the initial weight of the rocket. We can only note that, essentially, it does not save the situation since at sub-light velocities (in the order of 200,000–250,000 km/sec) the mass of the matter entering the rocket (inter-stellar matter is to be used for fuel) is negligible, while at velocities nearing that of light the efficiency of the motor would be negligible due to the very small difference between the velocities of the intaken and ejected masses.

proceed from the unknown, but he should not advance projects which dash head-on with the laws of physics which, we know for sure, will remain forever, if only as an approximation.

Let me clarify my idea on the example of the ambitious "magnetic bottle" project mentioned before. The idea is to store anti-matter in the shape of positrons (the second half of the fuel being electrons) in a wonderful "magnetic bottle".[§§] Let us not even consider the necessary

The author finally loses the last remnants of his equanimity.

amount of fuel. What is beyond all possibility is how to suspend a ton, even a kilogram, of positrons in a magnetic field. The electrostatic

---

[§§]Positrons and electrons are a favourite fuel for interstellar wanderers because their interaction is the only known reaction in which the probability of the reactants changing into gamma-quanta is unity. Protons reacting with anti-protons produce mesons, hence the rest mass of the reaction products is not zero, which it should be in the case of an ideal fuel.

repulsion of so many positrons would be so great that any magnetic field within some limits of reason would have to be greater than any known natural or artificial field by tens of orders. A more preposterous idea could hardly have been devised, but....

It is sometimes suggested, to save the situation, that isolated groups of positrons and electrons could be kept together if they are arranged like ions in a crystal lattice. There is no need to go into computations to appreciate the "revolutionary" nature of such a technological innovation. To me it seems more like a witch-doctor's incantations. With equal right we might expect man to evolve lungs powerful enough for him to blow a railway train to a standstill (and this is not an exaggeration!).

If we want to dream of storing anti-matter, why not let our fancy run away completely? Why not imagine that the manufacture of anti-atoms has progressed so far that we can produce the reciprocals of all the elements of the periodic system? Out of these anti-atoms we make an anti-metal (anti-iron, for example) which can easily be suspended in a magnetic field.¶¶ Of course, we know a safe way of dividing anti-atoms into positrons and anti-protons. We also know how to make the positrons react with the electrons. True, we have anti-protons left on our hands. They, too, must be turned into quanta. Well, why not suppose that we have learned to control the proton-anti-proton reaction and that the reacting particles inevitably turn

¶¶A good conductor can be suspended in an electro-magnetic field (the famous "Mohammed's coffin" trick), and super-conductors are even better. Well, we can store our anti-iron in a super-conducting state.

into quanta. All this phantasmagoria has one advantage over the positron-storage project: it contains nothing contradicting common sense, nothing which could be branded as absurdity pure and simple.

I am very proud of my project and think that it is worthy of being ranked alongside all the others. Unfortunately, I am unable to offer an equally successful project of a motor for a rocket weighing 10^5 tons which would be capable of accelerating it to sub-relativistic velocities without vapourising it. If I could, I might even find myself in the photon camp.

But if we wish to speak seriously and also bring in the psychological side of the problem, the idea of photon rockets is not so much an indication of the boldness of human fantasy as of its hideboundness. Even in their fancies people are unable to conjecture things without a point of departure in contemporary science and technology. And that is why they follow the beaten, but utterly wrong, path of hypertrophic expansion of known methods.

This is a tendency of all myths and fairy-tales. Helios, the Sun-god of the ancient Greeks, travelled across the sky in a chariot drawn by a splendid brace of horses. The horses were really fine, their pace was tremendous, sometimes they were even furnished with wings. But there is no "fresh technological idea" in the tale. Simply the gods could get the best of horses from the stables of Mt. Olympus. The Greeks dreamed of wonderful horses and chariots without suspecting the much more wonderful power of steam.

In the age of steam, awe-inspiring steam coaches were conjectured, though internal

A historico-mythological digression.

combustion engines were practically around the corner.

The ancients sent their heroes to the Moon on the backs of huge birds. Edgar Poe, in the 19th century, used a balloon for the same purpose.

The beginning of the 20th century abounded in "electrically-inclined" fantasts until atomic energy became known. In our day of rockets and nuclear energy it is only natural that photon ships should have been conceived. It seems very likely that their fate will be that of their predecessors. If the stars are ever reached, the trip will be carried out by means which we could never have dreamed of.

There is no need to overtax our rockets. Exploration of the solar system is only just beginning, and they are good enough for the purpose. As for other stars and galaxies....

Several centuries ago the Frenchman Cyrano de Bergerac wrote a book about a flight to the Moon. Among the dozen or so extravagant modes of travel flippantly suggested by him, among the amusing gibberish, by sheer chance there slipped the idea of reaching the Moon with the help of a fire-works rocket. To do de Bergerac credit, it should be said that he cared little what emerged from under his pen and placed no stock in his projects. Nevertheless, he had suggested a rocket motor using the fairly modern fuel of gunpowder. By sheer chance he guessed correctly.

Our position is worse for it is just about certain that photon rockets with nuclear fuel are unsuitable, even for reaching the closest stellar system. To be sure, I personally cannot suggest anything better. It is always difficult to imagine the unimaginable, but in any case photon rockets won't take off for the stars.

In conclusion, allow me to answer a question which may have arisen in reading this chapter. What good is served by so viciously attacking photon rockets? After all, the dream of reaching the stars is wonderful.

The author apologises.

Yes, the dream is wonderful, but our dreams should be within reason. I believe that the authors describing photon vehicles do so with the best of intentions. The result, however, is to mislead people who are unable to verify their statements and must take them for granted. This, in turn, leads to statements that stellar flight is a matter of the next few decades, or to serious debates around such "problems" as the merits and demerits of motors working on high-frequency and high-energy quanta (light quanta) or on low-energy electromagnetic field quanta.

It is not merely a matter of distracting attention from unsolved but feasible problems. After all, no one is barred from dreaming, and dreams of stellar flight can only be welcomed. But people should not discredit physics by creating the impression that the principles are apparent and that it is almost time to get down to the drawing boards. Science does not need publicity stunts. From our present knowledge flight to the stars is as unthinkable as television or radio in Galileo's time. If a physicist, some contemporary of Galileo's, were asked what he thought of the possibility of transmitting an image over a distance of several hundred kilometres, he would, most naturally, reply: "Our knowledge of physics says that this is impossible. A miracle is necessary, a discovery which would alter all our conceptions of the world, something unimaginable and contrary — not to the laws of nature, of course — but to our knowledge of those laws."

And last.

The same goes for stellar flight. A whole complex of miracles is needed. It is foolish and naive to try and guess when and whether any future discoveries will at least give promise of flights into the universe, just as it is foolish to guess what such discoveries may comprise. We can only trust that, having worked so many miracles before, mankind will work this one as well. But no nuclear fuels, no super-, ultra- or extra-materials, no photon rockets will solve the problem.

Something Unknown is needed, as unknown as atomic energy to Neanderthal man. This naive, foolish, childish and unquenchable belief in the Unknown is as dear to me as to most people of our time.

Index

Printed in the United States
by Baker & Taylor Publisher Services